Sensational Springtime

Principal Authors

Kay Kent
Barbara Aston
Myrna Mitchell
Barbara Ann Novelli
Michelle Pauls

Contributing Authors

Angela Boese
Sheldon Erickson
Janis Fehlig
Paula Freidson
Carol Gossett

Ann Lewis
Alice Potts
Rhonda Shook
Michele Strayer

Richard Thiessen
Cheryl Vaughan
Virginia Welsh
Jim Wilson

Illustrator

Dawn Don Diego

Editors

Michelle Pauls
Betty Cordel

Desktop Publisher

Tanya Adams

Education Foundation

This book contains materials developed by the AIMS Education Foundation. **AIMS** (**A**ctivities **I**ntegrating **M**athematics and **S**cience) began in 1981 with a grant from the National Science Foundation. The non-profit AIMS Education Foundation publishes hands-on instructional materials (books and the quarterly magazine) that integrate curricular disciplines such as mathematics, science, language arts, and social studies. The Foundation sponsors a national program of professional development through which educators may gain both an understanding of the AIMS philosophy and expertise in teaching by integrated, hands-on methods.

ISBN: 978-1-932093-39-1

Printed in the United States of America

I Hear and I Forget,

I See and I Remember,

I Do, and I Understand.

-Chinese Proverb

Sensational Springtime

Table of Contents

Sprouting on Sponges

Topic
Seed growth

Key Questions
1. How long will it take for our seeds to sprout?
2. Which seeds will sprout first?

Learning Goals
Students will:
- observe seed germination, and
- keep a record of how long the different kinds of seeds take to germinate.

Guiding Documents
Project 2061 Benchmarks
- *Describing things as accurately as possible is important in science because it enables people to compare their observations with those of others.*
- *Change is something that happens to many things.*

NRC Standards
- *Employ simple equipment and tools to gather data and extend the senses.*
- *Plants and animals have life cycles that include being born, developing into adults, reproducing and eventually dying. The details of this life cycle are different for different organisms.*

NCTM Standards 2000*
- *Create and use representations to organize, record, and communicate mathematical ideas*
- *Recognize the attributes of length, volume, weight, area, and time*

Math
Graphing

Science
Life science
   seed growth

Integrated Processes
Observing
Predicting
Collecting and recording data
Interpreting data
Communicating

Materials
For the class:
   sponge squares (see *Management 4*)
   three or four types of seeds (see *Management 1*)
   containers for gardens (see *Management 5*)
   prediction graph (see *Management 6*)

For each group:
   calendar page
   sponge recording page

Background Information
In order to germinate, seeds must absorb water until they swell and burst their seed coats. Inside the seed is a tiny embryo, surrounded by stored food. The root is usually the first structure to emerge. It grows rapidly, absorbing water and minerals from the soil. Then the shoot begins to grow, pushing its way through the surface of the soil into sunlight. When the first leaves form, the plant begins to manufacture its own food.

The seeds suggested in this activity grow rapidly. When the seeds are kept moist, they germinate, and roots, stems, and leaves are soon visible. As they grow on the sponge, they will soon need more nutrients than they are receiving. Adding plant food to the water can prolong the plants' lives.

Management
1. Collect a variety of seed types—mustard, alfalfa, radish, mung bean, birdseed, etc. If desired, each student or group of students can take responsibility for a different type of seed and compare the growth. For best results, soak the seeds for several hours before "planting" them on the sponges.
2. This activity is best done in pairs or small groups and should be continued over several days or weeks.
3. Beforehand, if possible, observe and discuss a number of different kinds of seeds. Help the students to understand the purpose of seeds and the process of germination.
4. Select sponges with large pores. The fluffy car wash type works well. Cut the sponges into rectangles about five centimeters by eight centimeters (two by three inches). It is VERY important that the sponges be no thicker than about two centimeters (.75 inch). This will allow the top of the sponge (where the seeds will be sitting) to remain moist. Each group needs one sponge.

5. Collect containers in which to put the sponge gardens. The containers must be large enough to hold the sponges and be at least four centimeters (one inch) deep. The containers should also allow direct sunlight to reach the seeds. Pie tins work well. A jelly-roll pan could be used to hold multiple sponge gardens.

6. Prepare a graph for predicting (guessing) which seed will sprout first using a sheet of chart paper. Graphing markers are provided.

7. Each group will need a copy of the calendar page and the recording page. Before making copies, write in the month and dates on the calendar page.

Procedure

1. Review with students what they know about seeds and discuss the *Key Questions*. After the discussion, tell students they will be observing what changes occur in seeds as they germinate, how long these changes take, and which seeds will germinate fastest.

2. Assist students in preparing their sponge gardens. Each sponge must first be soaked with water until it is wet through, then placed in its container. Once it is placed in the container, the seeds can be placed on its surface and into its pores. Then the container should be filled with water until the sponge floats.

3. Give each group a limited number of seeds (no more than 10) so that they will be able to make accurate pictures of their sponge gardens.

4. Distribute the calendar page and sponge recording page to each group. Have students draw pictures of their sponge gardens. Be sure they draw the seeds, and help them label the seeds on their drawings. This will help them identify each seed as they observe the germination.

5. Ask the students to predict (guess) which type of seed will sprout or start to grow first. Have them put a marker on the graph to record their predictions.

6. Using the calendar, ask students what day they think their first seed will sprout. Have each student write his or her name in the box under the day.

7. Different group members should take responsibility for sketching the seeds and recording the sprouting predictions on the calendar.

8. Water the sponge gardens every day by adding water to the containers. Do not pour water directly onto the sponges, as this may cause the seeds to move and/or fall into the container. The tops of the sponges should be moist at all times. If they dry out, the seeds will not germinate.

9. Once germination is observed, the students will need to indicate what day the seeds actually grew. Use the calendar record sheet to record findings. Use markers to indicate which seeds began on which days.

10. When all seeds have germinated, gather the class together to discuss the results of their study.

Connecting Learning

1. What happened when you put your seeds on the wet sponge?
2. When did the seeds first germinate?
3. Which seed did you think would germinate first?
4. Which seed actually germinated first? ...last? Did they all germinate? Did all groups have the same results?
5. Did all of the seeds look the same as they began to grow? How were they alike? ...different? Explain.
6. What changes did you notice in your seeds as you watched them over time? What surprised you?
7. If you were going to set up another sponge garden, what kind of seeds would you like to try on it?
8. If you were going to grow another garden, what else could you use besides a sponge?
9. What would happen if we didn't use a wet sponge? What would happen to the seeds if your sponge got dry?
10. What are you wondering now?

Extensions

1. Use different kinds of seeds to see if they germinate in about the same time and in about the same way.
2. Use Unifix cubes to measure the growth of the seeds and record their heights.
3. Grow the seeds on something other than sponges—paper towels, cloth, etc. Compare the results.
4. Use different amounts of water on different sponges and observe the results.
5. Transplant some of the seeds to soil and keep the plants growing. Keep a journal of the changes they go through over time.

* Reprinted with permission from *Principles and Standards for School Mathematics*, 2000 by the National Council of Teachers of Mathematics. All rights reserved.

Sprouting on Sponges

_____ Month

Sunday	Monday	Tuesday	Wednesday	Thursday	Friday	Saturday

Sprouting on Sponges

Draw a picture of the seeds that you planted.

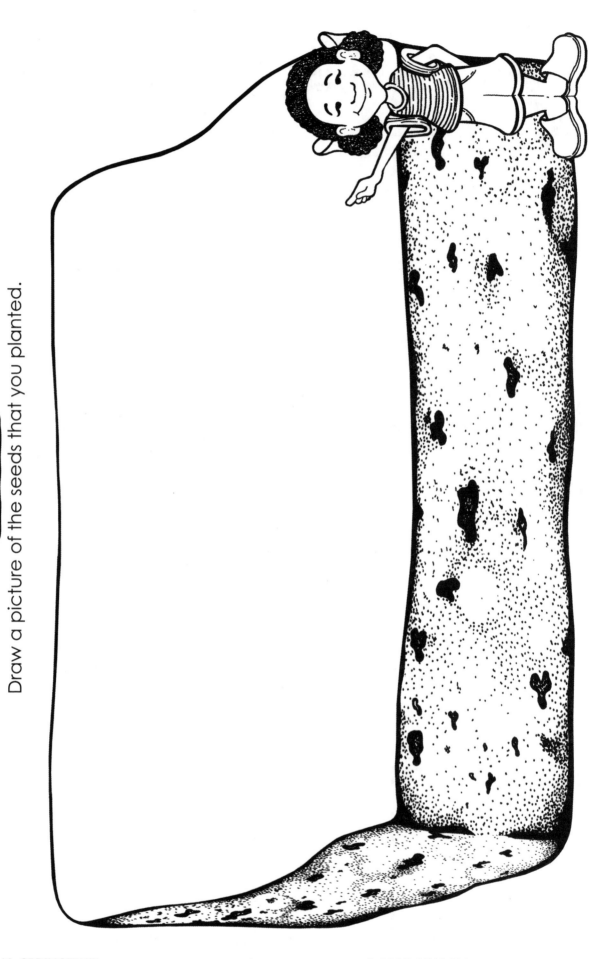

Sprouting on Sponges

Graph Markers

Sprouting on Sponges

I started my garden	I started my garden	I started my garden	I started my garden	I started my garden	I started my garden	I started my garden
added water	added water	added water	added water	added water	added water	added water
added water	added water	added water	added water	added water	added water	added water
added water	added water	added water	added water	added water	added water	added water
Yippee! Is that a sprout I see?	Yippee! Is that a sprout I see?	Yippee! Is that a sprout I see?	Yippee! Is that a sprout I see?	Yippee! Is that a sprout I see?	Yippee! Is that a sprout I see?	Yippee! Is that a sprout I see?
What a hoot! I see a root.	What a hoot! I see a root.	What a hoot! I see a root.	What a hoot! I see a root.	What a hoot! I see a root.	What a hoot! I see a root.	What a hoot! I see a root.
A leaf of green is what I've seen.	A leaf of green is what I've seen.	A leaf of green is what I've seen.	A leaf of green is what I've seen.	A leaf of green is what I've seen.	A leaf of green is what I've seen.	A leaf of green is what I've seen.

Topic
Seed growth

Key Questions
1. What will happen to wheat seeds when they are planted in an eggshell?
2. What will the wheat look like as it grows?
3. How fast will the wheat grow?

Learning Goals
Students will:
- plant wheat in eggshells and observe changes,
- use journals to record their observations, and
- chart the growth of seeds.

Guiding Documents
Project 2061 Benchmarks
- *Describing things as accurately as possible is important in science because it enables people to compare their observations with those of others.*
- *A lot can be learned about plants and animals by observing them closely, but care must be taken to know the needs of living things and how to provide for them in the classroom.*
- *Most living things need water, food, and air.*

NRC Standards
- *Organisms have basic needs. For example, animals need air, water, and food; plants require air, water, nutrients, and light. Organisms can survive only in environments in which their needs can be met. The world has many different environments, and distinct environments support the life of different types or organisms.*
- *Plants and animals have life cycles that include being born, developing into adults, reproducing, and eventually dying. The details of this life cycle are different for different organisms.*

NCTM Standards 2000*
- *Understand how to measure using nonstandard and standard units*
- *Use tools to measure*
- *Represent data using concrete objects, pictures, and graphs*
- *Recognize and apply mathematics in contexts outside of mathematics*

Math
Measurement
Graphing

Science
Life science
   seed growth

Integrated Processes
Observing
Comparing and contrasting
Communicating
Predicting
Collecting and recording data

Materials
For each student:
   two eggshell halves
   two sections of egg carton
   plastic teaspoon
   student pages

For the class:
   eyedroppers
   large bucket
   potting soil (large bag)
   wheat seeds (approx. 1 lb.)
   linear measuring tools (see *Management 6*)

Background Information
Wheat seeds grow rapidly. Students will discover that each seed grows into a specific type of plant. Wheat seeds will sprout when sprinkled on top of moistened soil. The germination of seeds is dependent on many factors such as light, amount of moisture, type of soil, and time.

Management
1. Prior to the investigation, ask parents to collect and send in an eggshell divided into two parts, and an egg carton, if possible. Save some extra eggshells and cartons for students who haven't brought any.
2. Wheat seeds are usually available in feed stores or health food stores.
3. Moistening the potting soil in a large bucket ahead of time helps students in the planting process.

4. Use eyedroppers as watering tools to discourage over-watering of wheat seeds.

5. Cut the egg cartons into two-section parts prior to the activity so that they are ready to receive the eggshell halves.

6. Provide appropriate measurement tools such as rulers or centicubes that will allow students to measure in centimeters.

7. To make a journal for each student, copy the cover page and the recording page back to back and fold in half. Make additonal copies of the recording page as necessary and nest them inside. Staple along the outside edge if desired.

8. Be sure you have a sunny location in your room where the egshells can be set to grow.

Procedure

Part One

1. Review previous experiences that students have had observing changes in seeds. Discuss what is needed to help seeds grow. Focus students' attention on seeds and changes in seeds by reading a non-fiction book about seeds. (See *Curriculum Correlation*.)

2. Ask the *Key Questions* and state the *Learning Goals*.

3. Demonstrate how to prepare the eggshells by scooping some potting soil into them and placing them in the egg carton sections.

4. Distribute the eggshells and cartons to each student. Write each student's name on his or her egg carton section. Have students fill their eggshells with potting soil and put them in the cartons.

5. Demonstrate how to sprinkle the wheat seeds over the soil using a plastic spoon and then carefully water them using the eyedropper.

6. Have students sprinkle wheat seeds on their eggs and water them.

7. Show students where they will be keeping their eggshell gardens, and have them move the gardens to that spot.

8. Distribute a journal to each student. Show them how to record their names and the beginning date, and explain how the journal will be used throughout the activity. Collect and store journals for future use.

Part Two

1. Once germination is observed, show students how to use the ruler or centicubes to measure the growth and record the measurement on the graph page.

2. Have students record their observations in their journals.

3. When all of the seeds have germinated, gather the class together to discuss the results of their study.

Connecting Learning

1. What changes happened in your seeds?
2. What happened first when you planted your wheat?
3. Did all of the eggheads grow at the same time? How were they different?
4. What else could we grow in our eggshells?
5. Can you think of a different container to grow plants in?
6. How would you change the experiment if you were going to do it again?

Extension

Have students make a "wheat watcher" by gluing together two cotton balls and making eyes and a beak from construction paper.

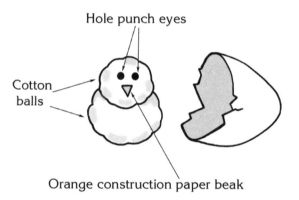

Hole punch eyes

Cotton balls

Orange construction paper beak

Curriculum Correlation

Cole, Joanna. *The Magic School Bus Plants Seeds: A Book About How Living Things Grow*. Scholastic, Inc. New York. 1995.

Gibbons, Gail. *From Seed to Plant*. Holiday House. New York. 1993.

Hickman, Pamela. *A Seed Grows: My First Look at a Plant's Life Cycle*. Kids Can Press. Toronto, Ontario. 1997.

Jordan, Helene J. *How a Seed Grows*. HarperCollins. New York. 1992.

* Reprinted with permission from *Principles and Standards for School Mathematics*, 2000 by the National Council of Teachers of Mathematics. All rights reserved.

Eggheads
and
Wheat Watchers

Wheat Watcher

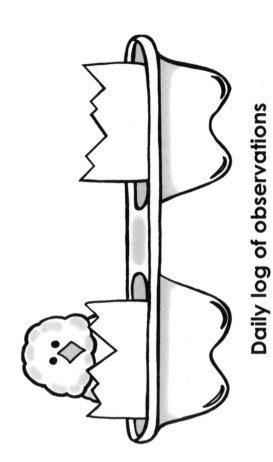

Daily log of observations

Wheat Watcher Productions

School _____

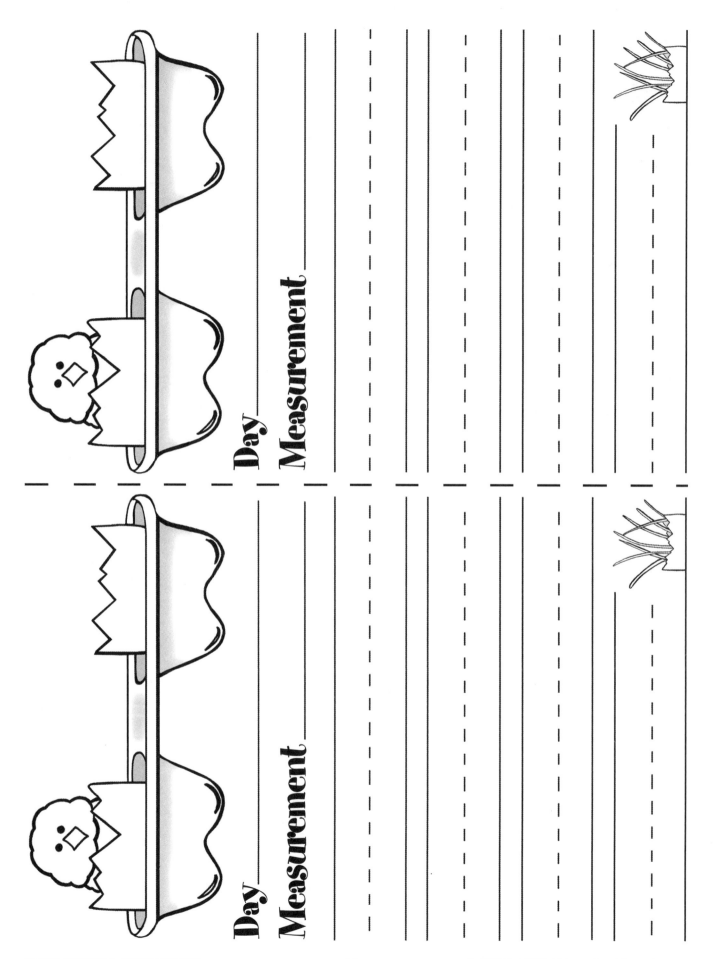

Day _____

Measurement _____

Day _____

Measurement _____

10

Eggheads and Wheat Watchers

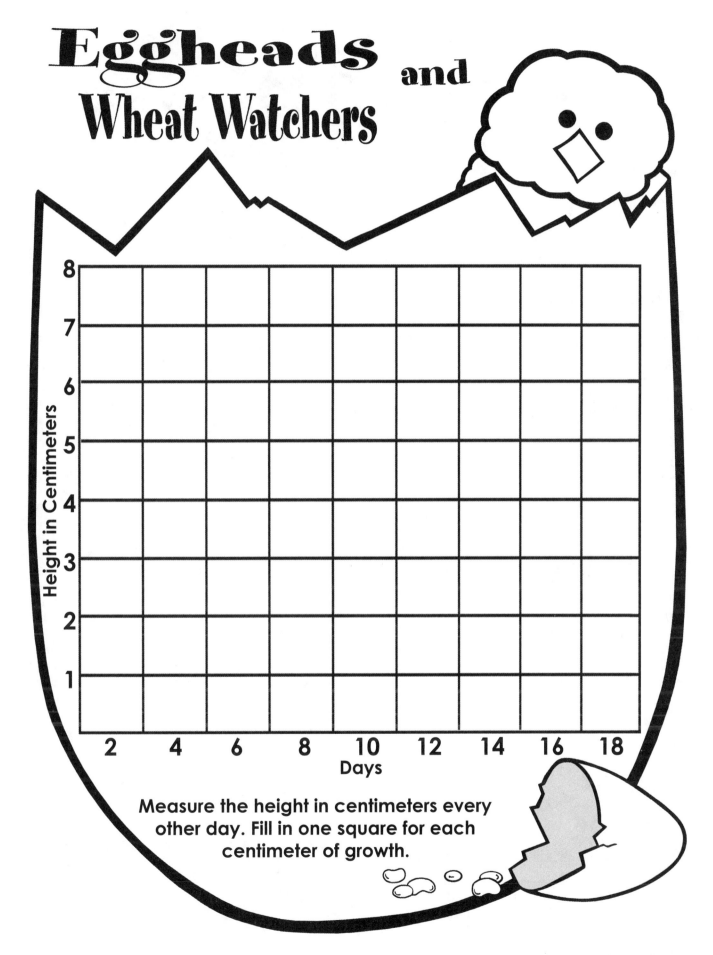

Height in Centimeters

8
7
6
5
4
3
2
1

2   4   6   8   10   12   14   16   18
Days

Measure the height in centimeters every other day. Fill in one square for each centimeter of growth.

11

Show it GROW

Topic
Plants: bulbs

Key Question
How can we show the growth of a bulb by observing it over time?

Learning Goal
Students will observe, describe, and graph the physical changes that occur in a bulb over several days and weeks.

Guiding Documents
Project 2061 Benchmark
* *Describing things as accurately as possible is important in science because it enables people to compare their observations with those of others.*

NRC Standard
* *Objects have many observable properties, including size, weight, shape, color, temperature, and the ability to react with other substances. Those properties can be measured using tools such as rulers, balances, and thermometers.*

NCTM Standards 2000*
* *Use tools to measure*
* *Measure with multiple copies of units of the same size, such as paper clips laid end to end*
* *Pose questions and gather data about themselves and their surroundings*
* *Represent data using concrete objects, pictures, and graphs*
* *Describe parts of the data and the set of data as a whole to determine what the data show*

Math
Measurement
    linear
Graphing

Science
Life science
    plant growth
    change over time

Integrated Processes
Observing
Comparing and contrasting
Predicting
Communicating
Collecting and recording data
Interpreting data

Materials
Bulbs (daffodil, paper white, amaryllis, etc.)
9-oz clear plastic cup
Small rocks
Unifix cubes
Unifix cube graph (see *Management 3*)
Permanent marking pen
Hand lenses, optional

Background Information
A bulb such as a daffodil bulb is actually an enormous bud with short central stems surrounded by thick leaves that store food and water. The bulb forms roots at the base of the stem. During the growing season, a bulb's central bud section sends up a shoot that produces a stem, leaves, and flowers above the ground.

After flowering, food for the next season is manufactured in the foliage and stored in the fleshy, underground leaves. Parts of the plant above ground die. The outer scales form a dry, papery covering. At the beginning of the next growing season, the bulb sprouts and flowers.

In this activity, students will graph the height of the plant as it grows from the bulb. By connecting the data points with a broken line, they will be able to record the change in height over time. Students will also keep track of the height of the plant using Unifix cube units. The line graph, the measuring (using Unifix cube units) of the plant's height, and the bar graph made of Unifix cubes will reinforce the observation of growth and change over time.

Caution: Daffodil bulbs are poisonous if eaten. The Poison Control Center reports that an active ingredient is concentrated in the bulbs that when ingested can cause vomiting. In larger amounts, paralysis may occur. Check before using any bulbs and warn students to thoroughly wash their hands after handling the bulbs.

Management

1. This is a two-part activity. In *Part One* students observe and record the growth of a bulb during a whole-class activity. In *Part Two* of the activity, groups will use their own bulbs and 9-oz cups to repeat the activity and record the results.

2. Beforehand, if possible, visit a nursery or have personnel from a nursery visit your class. Make arrangements for bulbs to be set aside for students to choose and purchase. If this is not possible, buy bulbs and allow children to select from this collection. If appropriate, ask for a donation from parents to defray the cost of bulbs. Sometimes local business or nurseries will donate the bulbs. You may want to use a variety of bulbs.

3. The broken line graphs will be done on chart paper. To prepare the graphs for both *Part One* and *Part Two*, position the paper so that the long side forms the horizontal axis. Use a marker to draw a horizontal line approximately 7.5 cm (3 inches) from the bottom edge of the chart paper. When students begin to graph the plant's growth, they will position the chart paper so that this line is at the top of the bulb. Also mark columns on the chart paper the width of the plastic cup. Encourage the class to decide on the appropriate question to use to title the graph.

4. Each group will need approximately half a cup of small rocks to place in the bottoms of the cups.

5. Make multiple copies of the Unifix cube page. It will be left intact and used for the bar graph, but it will also be cut apart so that students can glue Unifix cube trains to their chart paper graph to show the heights of their plants.

6. Each group will need a picture of a cup and bulb to glue on their chart paper graph (underneath the horizontal line you drew) for each day that they plot data on their line graph. They will record the date of the observation on the picture.

Procedure

Part One

1. Show students the bulb you will be using for the whole-class portion of the lesson.

2. Lead the class in a discussion of the *Key Question* and discuss how the investigation will be set up for the class portion.

3. "Plant" the bulb in the plastic cup by placing it large end down slightly into the rocks.

4. Add water to the cup until the rocks and large end of the bulb are just covered.

5. Put the cup on a table, ledge, or counter, where it can remain for an extended period of time.

6. When the cup with the bulb is in place, attach the graph to a wall or window behind it, aligning the horizontal axis with the top of the bulb. The graph will remain in place while the cup is moved along the table or ledge from column to column.

7. To demonstrate the recording procedure, position the cup in the first column. Discuss where the top of the bulb is and that there is no plant going up the column yet. (All measurements will be made from the top of the bulb.) Use a marker to make the first data point at the top of the bulb. Ask the students when they think they will notice a change in the bulb. Record the starting date on a picture of the cup and glue it under the horizontal line.

8. Although it will take several days for the bulb to begin to show its growth, once it begins, collect data on a daily basis. Be sure to record the date of the recorded observation on a picture of the cup.

9. Once the plant's growth is observed and a data point is plotted on the graph, use Unifix cubes to show students how to measure its growth from the top of the bulb. Cut out a Unifix cube strip and trim it to represent the number of Unifix cubes used to measure the plant's height. Glue the trimmed strip to the graph. As data points are added to the graph, connect them with a broken line.

10. Discuss the changes in the bulb and in the height of the plant above the bulb. While continuing to observe the class bulb each day, you may want to have the students begin *Part Two* of the activity.

Part Two

1. Once the students have had a chance to observe the changes in the class bulb example, they are ready to set up their own.

2. As a reminder, ask the students the *Key Question* and review the plotting of data points, the recording of dates, and the gluing of the Unifix cubes to the graph. Tell them that they will also make a bar graph by coloring in the Unifix cubes on a graph page to show the height of their plants.

3. Optional: Encourage students to examine their bulbs using hand lenses. Discuss the properties of the bulbs, similarities and differences, etc.

4. Have the students line the bottoms of their cups with small rocks and place their bulbs into the rocks. Direct them to add water until the rocks and the bottoms of the bulbs are just wet. Have them wash their hands after handling the bulbs.

5. Have students put their cups on the counter and help them tape their charts on the wall or windows behind them. Align the horizontal line with the top of the bulb. Direct the students to position their cups under the first column from the left.

6. Remind the students that they will need to slide their cups along the counter in front of their graphs when they start recording data. Review how to plot the data on the graph by making a dot at the top of the plant.

13

7. As soon as students have plotted the data on the graph, have them make a Unifix cube train the length of the plant's height. Invite them to color in the growth on the bar graph found on the student page. Have the students use a different color Unifix cube train for each measure. This will make the bar graph of colored cubes easier to compare.

8. Have students cut out a picture of a cup and bulb to attach to the bottom of the graph. Instruct them to record the date on the picture. Invite them to draw the bulb growth above the cup on the graph, making sure the pictures reflect growth up to the dots.

9. Direct students to record the heights of their plants by cutting a paper strip of Unifix cubes the same length as the real Unifix train and gluing the strip to the chart beside the bulb.

10. Encourage the students to continue to observe, measure, graph, and compare data over the next few days and weeks. Have students connect the data points on the chart paper graph with a broken line.

11. Urge students to keep their graphs so that when the plant begins to die they can compare/contrast and communicate their results.

9. How was the growth of your bulb like the growth of plants in your yard at home? How was it different from other plants you see?

10. What questions do you have about your plant or the class plant?

11. If you wanted to set up another investigation, what would you do? What questions would you ask?

12. Are there any other seeds or bulbs you would like to grow and graph their growth? If so, what are they?

Extensions
1. As a class, plant a bulb in soil and watch to see what happens. Make a graph and use Unifix cubes to show the results. Compare results with those bulbs that were planted in water.

2. Find the mass of the cup, the rocks, the water, and the bulb. Predict if the mass will change during the investigation. Talk about some of the reasons that there was or was not any change.

Curriculum Correlation
Bedard, Michael. *Emily*. Doubleday Books for Young Readers. New York. 1992.

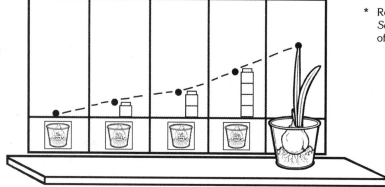

Connecting Learning
1. What part of the bulb grew?
2. When did the bulb first show growth?
3. How long did it take from the time we put the class bulb into the water and rocks for the first tip to show?
4. How did the time from the first column to the second column compare to the time between the columns after that?
5. What kind of a pattern do you see on your own graph? Is it the same as the line we saw on our class graph? How are they the same? …different?
6. Did all of our bulbs grow the same way? Did each of our plants measure the same number of Unifix cubes each day? Why or why not?
7. Why do you think the height of your plant was the same as or different from the height of someone else's? How did it compare with the growth of the class plant?
8. What did you discover about your plant as you watched it day by day? What things surprised you?

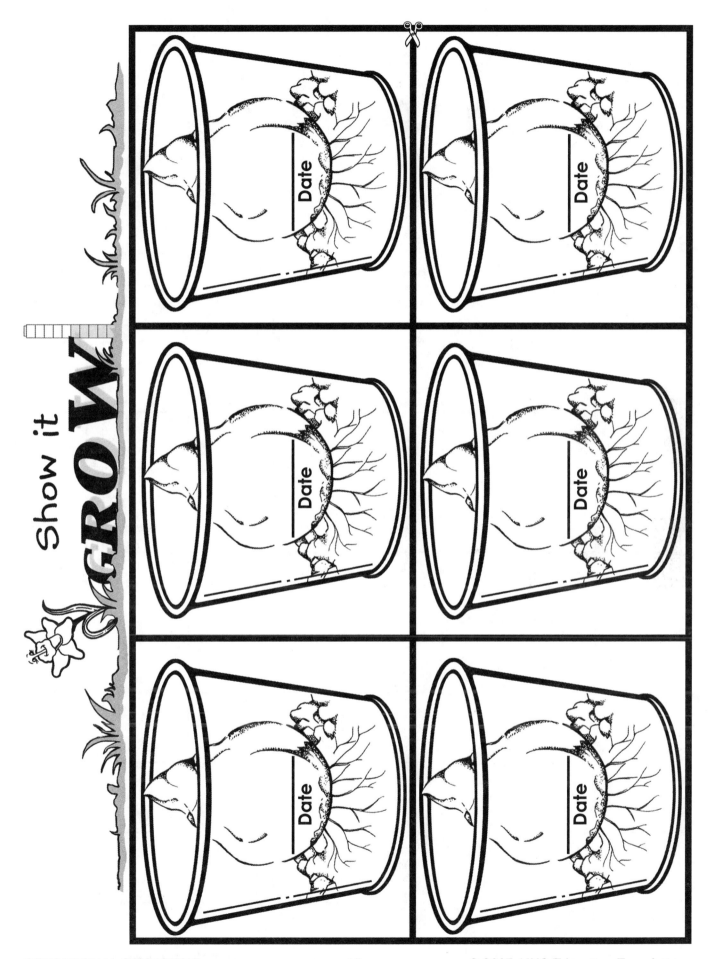

A Plant Patch

Topic
Plant growth

Key Questions
1. What do plants need to grow?
2. What do different vegetables look like as they grow?

Learning Goals
Students will:
- investigate the growth of various plants,
- observe change over time, and
- compare and contrast vegetables grown from seeds.

Guiding Documents
Project 2061 Benchmarks
- *Most living things need water, food, and air.*
- *Plants and animals both need to take in water, and animals need to take in food. In addition, plants need light.*
- *There is variation among individuals of one kind within a population.*

NRC Standards
- *Organisms have basic needs. For example, animals need air, water, and food; plants require air, water, nutrients, and light. Organisms can survive only in environments in which their needs can be met. The world has many different environments, and distinct environments support the life of different types of organisms.*
- *Employ simple equipment and tools to gather data and extend the senses.*

NCTM Standards 2000*
- *Measure with multiple copies of units of the same size, such as paper clips laid end to end*
- *Use tools to measure*
- *Pose questions and gather data about themselves and their surroundings*

Math
Measurement
    linear

Science
Life science
    plant growth

Integrated Processes
Observing
Predicting
Collecting and recording data
Comparing and contrasting
Interpreting data

Materials
Hard plastic wading pool (5-ft diameter, about 15 inches deep)
Charcoal briquettes
Rocks
Potting soil
Fast-growing vegetable seeds (e.g., radishes, lettuce)
Garden tools
Heavy string
Nails
Copy paper
Craft sticks
5-oz paper cups
Chenille stems
Unifix cubes
Hand lenses

Background Information
In order to germinate, a seed must absorb water until it swells and bursts its seed coat. Inside the seed is a tiny embryo, surrounded by stored food. The root is usually the first structure to emerge. It grows rapidly, absorbing water and minerals from the soil. Then the shoot begins to grow, pushing its way through the surface of the soil into sunlight. When the first leaves form, the plant begins to manufacture its own food.

When the seeds in this activity receive what they need, they begin to germinate and grow. It will take approximately 7-10 days for many vegetable seeds to begin to sprout. Students will be able to observe, record data, and measure the plants in the garden as they begin to grow.

Gardening allows children to nurture their plants while observing and working in their garden. This year-round activity is designed to allow children to care for their garden over time, observing patterns, while extending their curiosity and abilities in the natural world.

Management

1. This activity is divided into three parts. In *Part One,* students set up the garden and plan and plant a personal patch of garden. In *Part Two,* the students observe and measure different plants from their patches. In *Part Three,* students harvest and replant as needed.

2. Set up a growing corner in the classroom. This area should have a good natural light source or a grow light to provide light.

3. The wading pool should be placed in the growing area and a layer of rocks and charcoal should be placed to completely cover the bottom of the pool.

4. Cover the rocks and charcoal with the soil, filling the pool about two-thirds full.

5. Mark off the pool so that each child will have a section. Use string to make radiating spokes coming from the center of the pool. Use large nails to keep the string in place both in the center and at the outside edge.

6. Visit a nursery or garden department prior to the activity and buy the seeds, choosing ones that are in season and will grow quickly in the garden.

7. Check germination time so that children will be in school when the seeds begin to sprout. If carrot seeds take 7-10 days and you plant on Monday, the seeds should sprout during the following week.

8. Use craft sticks to mark the individual student plots. Use a permanent marker to put students' names and the date on the sticks.

9. Poke several small holes in the bottom of some of the 5-oz cups to use for watering. Use a chenille stem to make a handle loop on the top of each cup. Leave some cups without holes and loops.

10. Each student will need up to four *Seed Labels,* depending on how many different kinds of seeds he or she plants.

11. Prepare a journal for each student. Copy pages two and seven on the back of pages one and eight. Copy pages four and five onto the back of pages six and three. Assemble the journals and fold.

Procedure

Part One

1. Ask students what vegetables they have seen grow or what vegetables they like to eat. If possible, have some of the real vegetables available for students to see and taste.

2. Ask students what they know about how these vegetables grow.

3. Distribute the unopened seed packs and listen to the students as they discuss the plant pictures and prior knowledge of seeds.

4. Provide opportunity for students to observe each kind of seed that will be planted in the garden. Allow time for the students to talk about which seeds they will use.

5. Allow each child to choose the "patch" of garden he or she will plant. (These may be assigned if preferred.)

6. Discuss the needs of the seeds, including the space needed for the plants to grow.

7. Allow students time to discuss what they want to plant and where in their garden patch they want which plants to grow.

8. Distribute the journal and have students draw the plants they plan to grow and where they plan to plant the seeds.

9. Pour the seeds out onto individual sheets of paper. Tape the seed packet to the appropriate paper so students will know what seeds they have selected.

10. Have each student take eight to 10 seeds of one kind. Direct students to save one of the seeds they chose to tape into their journals.

11. Discuss with the students the planting requirements for each type seed. Model for them planting the seeds in one patch, the "class patch."

12. In small groups or in pairs, have the children go to the growing area and plant their first set of seeds. Caution them to cover their seeds lightly with soil.

13. Give each student a *Seed Label* and a craft stick. Have them draw a picture or write the name of the seed they just planted on the label. Direct them to glue or tape the label to a craft stick and place it next to the seeds they just planted.

14. Have the students again take eight to 10 seeds, tape one in their journal, plant and label the rest. Have them repeat this procedure as necessary for the size of the patch.

15. After all children have had a chance to plant, discuss the other things the seeds will need.

16. Show the children the cups they will use for watering and the correct way to water their patches. Explain that it is easy to overwater and underwater, and that they should check their patches each day.

17. Demonstrate for students how to use the cup with the handle on top and the holes in the bottom as the watering cup. Holding the cup by the handle, pour some water from a second cup (one without holes) into the watering cup. Move the watering cup slowly over the planted area to gently rain on the plot. Watering may be easier if children work in pairs.

18. Allow time each day for children to visit the garden and see if anything is growing in their patches.

19. When sprouts first appear, have the students draw pictures in their journals.

20. Allow time for students to continue to visit the patch, make observations, and draw pictures in their journals until the garden is ready to harvest.

Part Two

1. As plants become mature, it is time to harvest. Seeds will be ready to harvest at various times. Spinach takes 40-45 days, radishes are ready in 20-30, lettuce can be used in 40-60 days, etc. This part of the lesson will depend upon what is planted.
2. Gather the students around the garden and discuss what the garden looks like and how it has changed over the last few weeks.
3. Ask them to observe one plant in their patches and draw it in their journals.
4. Ask the students what they think the plant looks like under the ground. Instruct them to add the part of the plant underground that they can't see to their drawings.
5. After the students have drawn what they think the plant looks like underground, have them pull one plant out of their patch. Remind them to pull gently so that all parts, including the roots, will come up.
6. Use hand lenses to observe the plant. Discuss what the plant looks like, what they thought it would look like, and the ways that it is different. Have them place the real plant on the page of their journals and draw what the plant really looks like. Encourage the students to add as much detail as possible.
7. Have them use Unifix cubes to measure the parts of the plant. Tell them to color in the number of cubes needed to measure each part on the plant drawing.
8. Encourage students to compare the plant they pulled up (harvested) with the plants of other students. Which plant was the longest from leaf-tip to root-tip? Which had the longest leaves?
9. Have each student bring their plants to a table area. Ask them to put the plants in order from shortest to tallest. Allow time for them to manipulate the plants until they agree that they have the correct order.

Part Three

1. After the first harvest, invite the students to return to the garden and gather round.
2. Ask what other plants in their patch are similar to the plants that they already harvested. Ask which ones are different and discuss the differences.
3. Choose another plant to harvest. Carefully observe the way that it is growing. Have children share their ideas about the plants and what they will look and taste like.
4. Continue to garden and harvest, discussing and measuring and recording as they go. (Copy journal pages as needed.)
5. At the conclusion of this part of the garden experience, have the students write or draw a final summary of what they did with their patches and what the results were.

On-going Gardening

1. When the vegetables that were grown from seed are all harvested, ask the children what type of plants they would like to grow in the patch. Some suggestions: grains (seeds can be purchased from a health food store); bulbs (onions); tubers (potatoes); flowers from seeds, bulbs, or cuttings; or grass.
2. Keep the garden patch growing the entire school year.

Connecting Learning

1. What happened when you planted your seeds?
2. How many seeds did you plant? How many seeds sprouted? How many seeds (if any) did not sprout?
3. How did the different plants look after they grew?
4. What did your plants need to grow well?
5. What did you have to do to take care of your plants? How often did you have to water them? Did they all need the same amount of water? Explain.
6. What part of the plant did you see grow first?
7. How many days did it take until the first sprout appeared in your patch?
8. What was your biggest plant? What was your smallest?
9. If you could grow other plants in your garden, what would you grow?
10. Did any of your plants die? What do you think happened? Explain.
11. If you had a patch in a different section of the garden, do you think it would make any difference? Explain.
12. What are you wondering now?

Extensions

1. Read poems about plants, make posters of the poems, and post them on a stick in the center of the garden. As student learn the poems, replace with new poems.
2. Read *Tops and Bottoms* by Janet Stevens (Harcourt Brace & Company. New York. 1996). Compare the garden in the story with the students growing corner.
3. Send a few seeds home with students to plant. Compare the growing of the plants at home with the growing of the plants in their plant patch at school.
4. Use an Unifix cube train to measure other plants. Invite students to measure the plants that they find on the playground. Have them draw a picture of each plant they measure and record the number of cubes it took to measure the parts.

*   Reprinted with permission from *Principles and Standards for School Mathematics*, 2000 by the National Council of Teachers of Mathematics. All rights reserved.

A Plant Patch

Seed Labels

My seeds

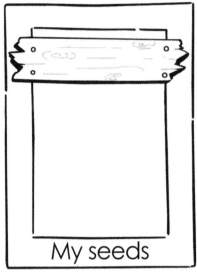

My seeds

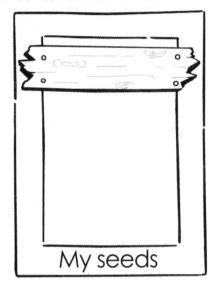

My seeds

My seeds

My seeds

My seeds

My seeds

My seeds

My seeds

My Patch Planner

Let me tell you about my plant patch.

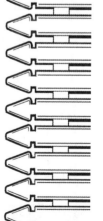

These are the seeds I used in my patch.

My seeds

My seeds

My seeds

My seeds

Draw and measure the whole plant.

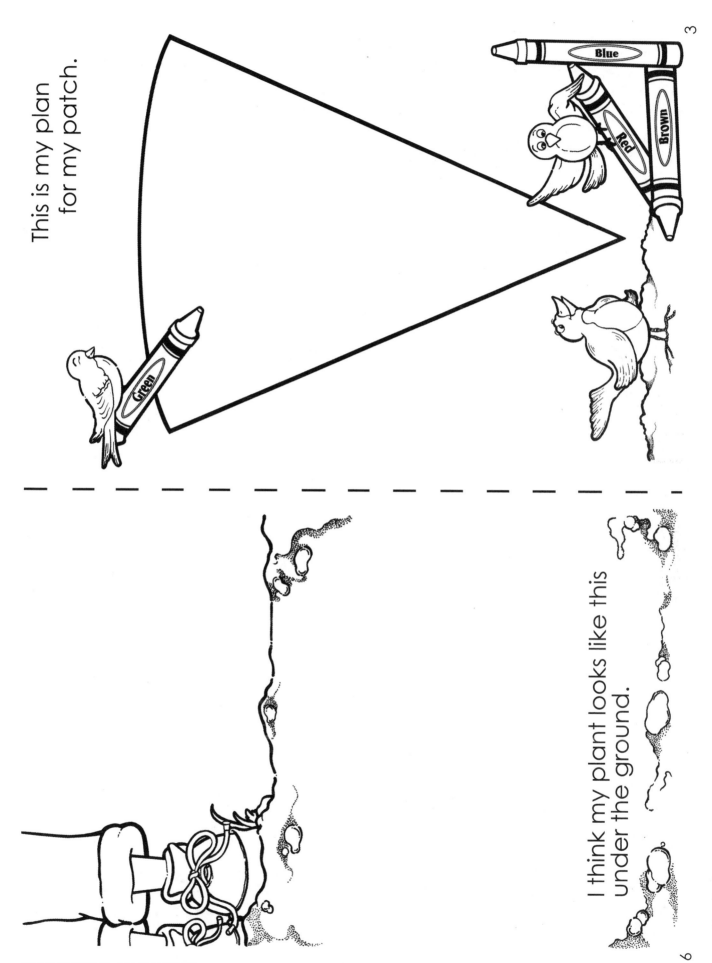

This is my plan for my patch.

Blue

Brown

Red

Green

I think my plant looks like this under the ground.

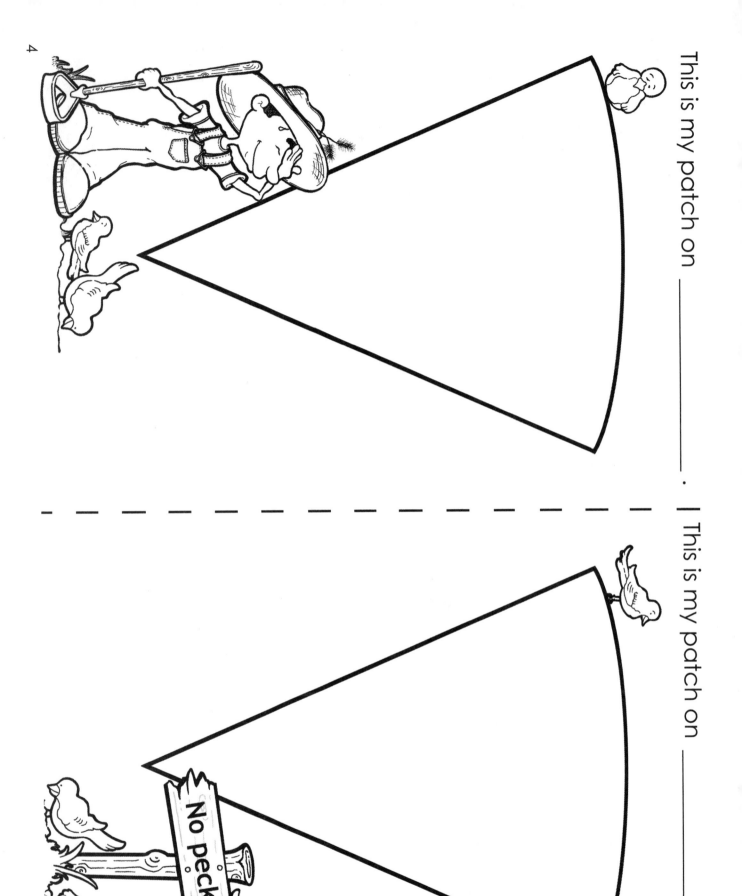

4

This is my patch on _____.

This is my patch on _____.

No Pecking

5

SENSATIONAL SPRINGTIME

24

© 2007 AIMS Education Foundation

A Seed Grows

Tune: Frère Jacques

Plant a se-ed. Plant a se-ed.

In the ground. In the ground.

Give it light and wa-ter. Give it light and wa-ter.

Watch it grow. Watch it grow.

First the roots grow.
First the roots grow.
Undergroud. Underground.
Then the stems and leaves grow.
Then the stems and leaves grow.
Toward the sun. Toward the sun.

It grows taller.
It grows taller.
Every day. Every day.
We can see it changing.
We can see it changing.
As it grows. As it grows.

Spot the Difference

Topic
Heredity

Key Questions
What picture of the plant or animal is different from the other two? How is it different?

Learning Goal
Students will understand that there are variations among individual plants and animals of the same kind.

Guiding Document
Project 2061 Benchmark
- *There is variation among individuals of one kind within a population.*

Science
Life science
heredity

Integrated Processes
Observing
Comparing and contrasting
Concluding

Materials
For each student:
student pages
crayons

Background Information
In nature, there are many different kinds of animals and plants that are easy to distinguish from one another. However, even animals and plants of the same kind show variations that let us tell them apart. The objective of this activity is to simply engage students in noticing and thinking about this idea. Students will look at four different sets of pictures, each showing three animals or leaves. They will be asked to find one of the three that is different from the other two. As an assessment, students are asked to draw three plants or animals, where two of them are the same and one has something about it that is different from the other two.

The following is information about the animals and plant that are used in this activity:

The Tiger Swallowtail Butterfly is yellow with black markings. Its color and markings are the reason for the its name. These butterflies have long "tails" on their back wings that make them look something like the long, pointed tails of swallows.

Ladybugs are almost always brightly colored, generally red, yellow, or orange, with black spots.

The Japanese maple tree has simple palmate leaves that look somewhat like the palm and fingers of a hand. The fingers of the leaf are called lobes. While most leaves on a Japanese maple will have five lobes, they can have as many as 11.

Management
1. This activity is best done as a whole-class activity where each student has a set of the pictures and has access to crayons.
2. Cut each of the student pages in half lengthwise so that each animal/plant is on its own strip.

Procedure
1. Tell the students that they will be looking at pictures of plants and animals. Tell them that in each picture one of the three animals or plants will have something about it that is different from the other two. Their job is to spot the difference.
2. Hand out the pictures of the three kittens. Tell them that these kittens are brothers and sisters and that they were all born on the same day. Ask the students if they can spot anything different about one of the kittens. [One of the kittens has more whiskers than the others.] Ask the students what they think would be a good name for this kitten. Invite the students to color the two kittens that look the same.
3. Distribute the pictures of the butterflies. Tell students that there is something different about one of butterflies. Ask them if they can spot the difference. [The markings on one of the butterflies are different from the other two.] Tell students that these butterflies are called Tiger Swallowtail Butterflies because of the black markings on their wings and because they are almost always yellow. Ask the students to use a yellow crayon to color the two butterflies that look the same.
4. Give students the pictures of the ladybugs. Ask them if they can find something different about one of the ladybugs. [The number of spots on one of the ladybugs is different from the other two.] Tell them that ladybugs are almost always either red, yellow, or orange. Ask students to use an red crayon to color the two ladybugs that are the same.

5. Give students the pictures of leaves from a Japanese maple tree. Ask them to compare the leaves to their hands. Do the leaves look a little bit like their hands? How are they the same? Is one of the leaves different than the other two? How is it different? Tell them that these leaves come from a Japanese maple tree and that most of the leaves have five "fingers" just like we do, but sometimes they will have more, like the one in the picture. Ask them to color the two leaves that are alike.

6. As an assessment, ask the students if they can draw three plants or animals of the same kind, where one of them has something about it that is different from the other two. Ask students to share their pictures with another student and to ask the other student if they can tell what is different in one of the objects that was drawn.

Connected Learning

1. Are all Tiger Swallowtail Butterflies the same? Explain.
2. Are all boys the same? How are they different? [different hair color, different height, different weight, hair is a different length, etc.]
3. How were the ladybugs the same? How were they different?
4. How are twins the same? How are they different?
5. Have you ever seen tulips? How are they the same? How are they different?

Extension

Ask two students to describe their pet cats. Would you be able to tell them apart? How would you know they are cats? [The have fur, four legs, whiskers, a tail, and they meow.] How are they different? [one is larger, one has a different pattern of spots, etc.]

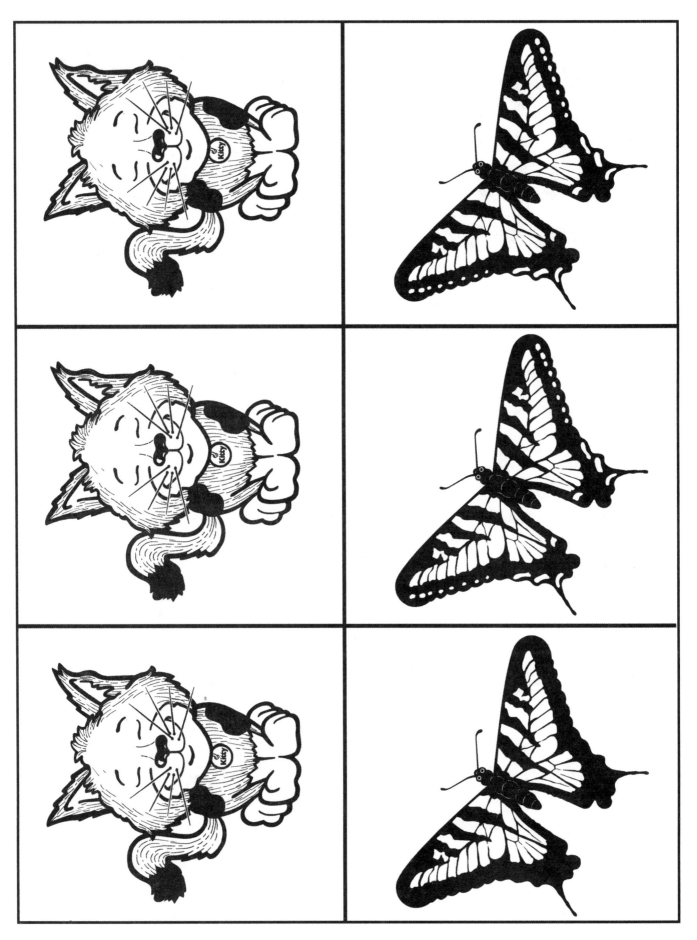

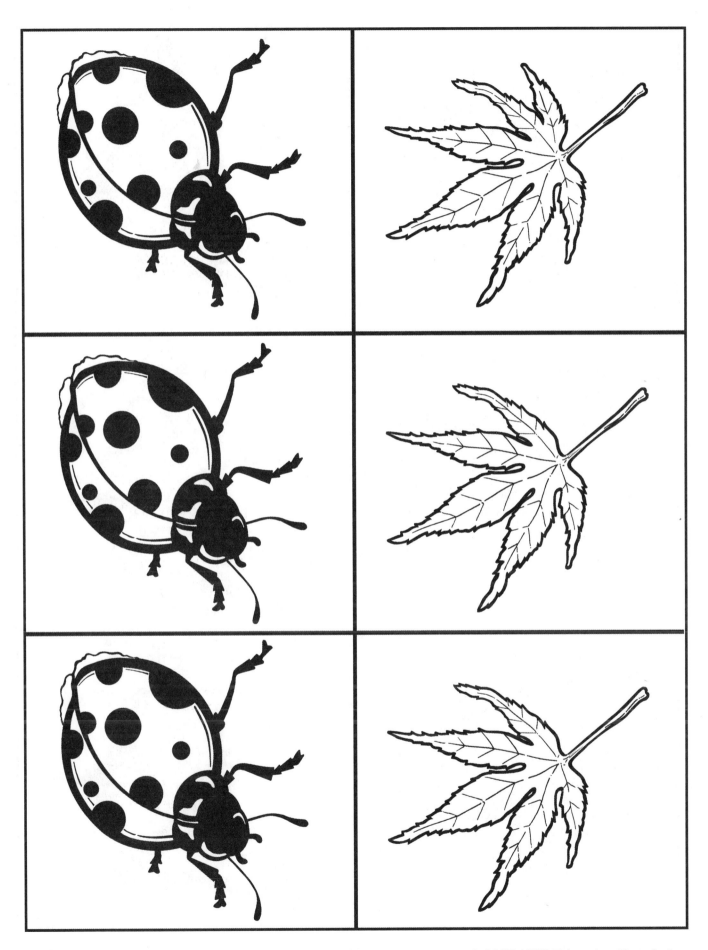

Find the Family

Topic
Plants

Key Question
How can you determine the parent of a plant?

Learning Goals
Students will:
- recognize the differences in plants, and
- identify plants of the same type by their common appearance.

Guiding Documents
Project 2061 Benchmarks
- *Some animals and plants are alike in the way they look and in the things they do, and others are very different from one another.*
- *Offspring are very much, but not exactly, like their parents and like one another.*

NRC Standard
- *Plants and animals closely resemble their parents.*

Science
Life science
   botany
      plants

Integrated Processes
Observing
Comparing and contrasting

Materials
Card set (16 cards)
Colored pencils or crayons

Background Information
Plants and animals closely resemble their parents. In plants, the difference is often only size. Mature plants will produce seeds or fruit. There are eight plants represented in the set: pumpkin, corn, iris, rose, cactus, date palm, oak tree, and pine tree. Each has one card showing a mature plant and one card showing a young plant.

Management
1. Multiple card sets will be needed for a class (one set for a pair of students), or one set may be used at a center.
2. Prepare the cards beforehand by copying them on card stock, cutting them out, and coloring them appropriately. If they are laminated, they will be more durable.

Procedure
1. Have the students work in pairs with a set for each pair.
2. Instruct the students to put all the cards face up in front of them.
3. Have the students compare the cards and pair up the parent and child pictures. When they have paired all of the pictures, have them make two rows of cards with the parent on top and the child underneath it.
4. When the students have the eight pairs in two rows, have them discuss how they know what plants are related and which of the pair is the child.
5. Instruct the students to turn all the cards face down. Have them mix the cards up and put them in a 4 by 4 array.
6. Choose a student to go first to turn over two cards. If a child/parent pair is made, the student keeps the pair. If no match is made, the cards are returned to the face down position.
7. Have the students alternate turns until all the cards are paired. The winner is the student with the most cards.
8. Play the game as often as necessary.

Connecting Learning
1. How can you tell if two plants are related (parent and child)? [similar appearance]
2. How is the parent of each pair different from the child? [bigger, often has flower or fruit]
3. What are you wondering now?

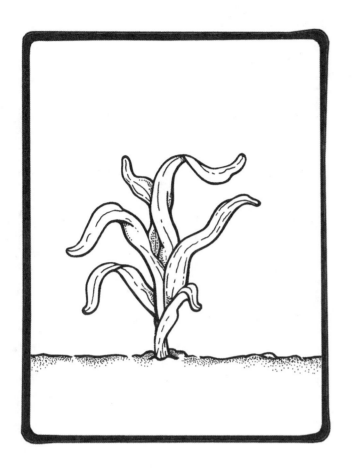

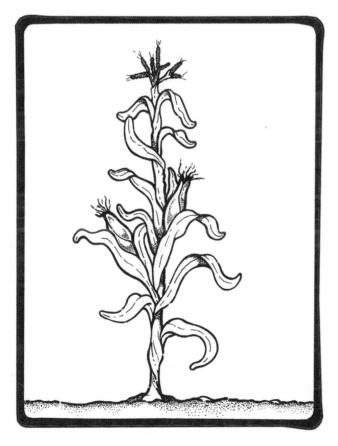

31

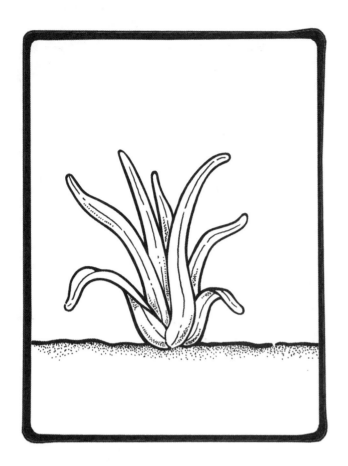

Who's My Mom?

Topic
Heredity

Key Question
How are offspring like their parents?

Learning Goals
Students will:
• identify and match animal offspring to their parents,
• recognize characteristics that distinguish one animal from another,
• create an imaginary animal and its offspring, and
• sort and graph animals according to their characteristics.

Guiding Documents
Project 2061 Benchmarks
• *Some animals and plants are alike in the way they look and in the things they do, and others are very different from one another.*
• *Offspring are very much, but not exactly, like their parents.*
• *A great variety of kinds of living things can be sorted into groups in many ways using various features to decide which things belong to which groups.*

NRC Standard
• *Plants and animals closely resemble their parents.*

NCTM Standard 2000*
• *Collect, organize and describe data*

Math
Graphing
    real graphs
Sorting
    Venn diagrams

Science
Life science
    animals
        heredity

Integrated Processes
Observing
Classifying
Comparing and contrasting
Recording data
Communicating

Materials
Part One:
    sticky notes

Part Two:
    animal books (see *Curriculum Correlation*)
    animal cards
    scissors
    grouping circles (see *Management 2*)

Part Three:
    animal construction materials (see *Management 7*)
    glue
    scissors
    tape
    round toothpicks

Background Information
Heredity is the "stuff" we inherit from our parents and ancestors. In humans and other animals, heredity is passed from parent to offspring by means of genes. Heredity determines things as mundane as eye color, and things as important as susceptibility to disease.

An easy way for primary children to study heredity is by examining how offspring resemble their parents. Because many physical characteristics are genetically determined, there tend to be many observable similarities between parents and their offspring.

Management
1. These activities can be done over a period of days or weeks, either as whole-class activities or in stations.
2. Each group will need two grouping circles to make a Venn diagram. Grouping circles are available from AIMS (order #4621), or you can make your own using two different colors of yarn. Cut the yarn to approximately 1.5 meters and tie the ends together to form a closed loop.
3. To support this activity, create a bulletin board with descriptions and pictures of animals and their offspring.
4. This activity does not get into traits that make children look like their parents such as eye color, skin color, etc. It focuses on broad categories such as having two arms, two legs, and 10 fingers like our parents.
5. You will need to copy one set of the animal cards for each group of students to use in *Part Two*.
6. The art project in *Part Three* can be done at an art station or with the entire class.

7. Provide a variety of materials for students to use when making their imaginary animals in *Part Three*. Large and small potatoes, oranges, bananas, or marshmallows could all serve as bodies; raisins, grapes, or peas could be used as eyes; baby carrots, paper clips, or construction paper could serve as legs, and so on.

8. Round toothpicks are suggested for use with the construction of the imaginary animals because they are less fragile than the flat ones.

Procedure
Part One
1. Ask students how they know whether something coming down the sidewalk is a person or an animal.

2. Invite students to tell you some characterisitics of humans. [Two legs, two arms, a head, two eyes, one nose, hair, two ears, 10 fingers, 10 toes, etc.]

3. Ask them to tell you the characteristics of a baby human. [same as above]

4. Tell them that they will be studying how offspring resemble their parents. Define offspring for them as babies. Baby dogs are called puppies. Baby cats are called kittens. Ask how baby dogs resemble their parents.

Part Two
1. Introduce *Part Two* with a video about animals and their offspring or read a story that has factual information (see *Curriculum Correlation*).

2. Have students get into groups and distribute one set of animal cards to each group. Have them cut out the cards and ask them to match the offspring to the parents by looking closely at the characteristics.

3. As a class, go over the correct matches (see *Solutions*) and discuss any animals that students had difficulty identifying. Pay special attention to animals that differ from their offspring. If desired, you may also take this time to share the different names that baby animals have.

4. Distribute the grouping circles and have groups set up one-circle Venn diagrams. Ask them to sort and classify the animal cards in various ways. Suggested sorts include size, coverings (feathers, scales, hair, shells), how they move, where they live, or other physical features.

5. Assist them by writing labels based on the characteristics they have chosen. One label should go inside the circle, and the other should go outside the circle. For example, lives in water/does not live in water.

6. If appropriate, move on to two-circle Venn diagrams.

Part Three
1. Using a variety of materials, have students construct an imaginary adult animal and its offspring.

2. Randomly place all of the completed models in a space on the floor and ask students to sit in a circle around the animals.

3. Call on a student to select an adult animal and find the matching offspring. (They may not select their own animals.) The student must tell why he or she thinks the offspring matches the adult. Identify the student who made the adult animal and ask if the classmate was correct. If the student was correct, the animals are placed to one side and another student is asked to select another set of adult and offspring animals, and so on. If a student does not make a correct match, then the animals are placed back with the others that have yet to be identified.

4. After all animals have been matched, have the students make a class graph of their animals using a variety of characteristics such as size (small, medium, large), construction materials (marshmallows, potatoes, toothpicks), or colors. This can be done as a floor or table graph.

5. Ask the students what information this graph gives them (most, least, equal amount, five more animals made of potatoes than animals made of marshmallows, etc.). Students may want to determine different ways to graph the animals and change the physical graph each day during the animal unit of study.

Connecting Learning
Part One
1. What are offspring? [babies]
2. In what ways do most offspring look like their parents?
3. What are some characteristics that an adult and an offspring might share?
4. How do you know that a puppy isn't the offspring of a fish?

Part Two
1. Which adult animal cards were the most difficult to pair with an offspring card? Why?
2. Discuss how your group sorted the cards. What were some of the headings you chose?
3. Share some interesting new facts you learned about an animal pictured on the cards.
4. What animals are you interested in learning more about now?

Part Three
1. Describe your imaginary animal. What characteristics did the adult and offspring share?
2. Why was it more difficult to match some adults with offspring than others?
3. How did your class graph the imaginary animals?
4. What are you wondering now?

Extensions

1. Play tic-tac-toe with the animal cards. Make a large tic-tac-toe grid and turn the cards into playing pieces by marking the back of each one with an "X" or an "O." Students take turns drawing cards (one player from the X stack, and one from the O stack). If they can correctly identify the animal, they get to play the card. If not, the other player gets to go. The first player to have three in a row horizontally, vertically, or diagonally wins.

2. Play a version of "Go Fish" using the animal cards, with the goal being to match adult and offspring cards.

3. Play a concentration game with the animal cards, matching the adult with the offspring, or combine sets to match adult to adult, or offspring to offspring (adult bear to adult bear, or tadpole to tadpole).

Curriculum Correlation

Literature

Bauer, Marion Dane. *My Mother is Mine*. Simon and Schuster. New York. 2001.

Eastman, P.D. *Are You My Mother?* Random House. New York. 1998.

Fisher, Aileen. *You Don't Look Like Your Mother*. Mondo Publishing. New York. 2002.

Guarino, Deborah. *Is Your Mama a Llama?* Scholastic, Inc. New York. 1989.

Keats, Ezra Jack. *Over in the Meadow*. Four Winds Press. New York. 1971.

Ryan, Pam Munoz. *A Pinky is a Baby Mouse*. Hyperion Books for Children. New York. 1997.

Tafuri, Nancy. *You Are Special, Little One*. Scholastic Press. New York. 2003.

Solutions

Here are the animals matched with their offspring. The correct names have also been given in each case.

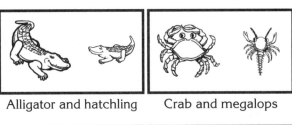

Alligator and hatchling

Crab and megalops

Dog and puppy

Flamingo and chick

Eagle and eaglet

Mouse and pinky

Polar bear and cub

Seal and pup

Penguin and chick

Frog and tadpole

Whale and calf

Butterfly and caterpillar

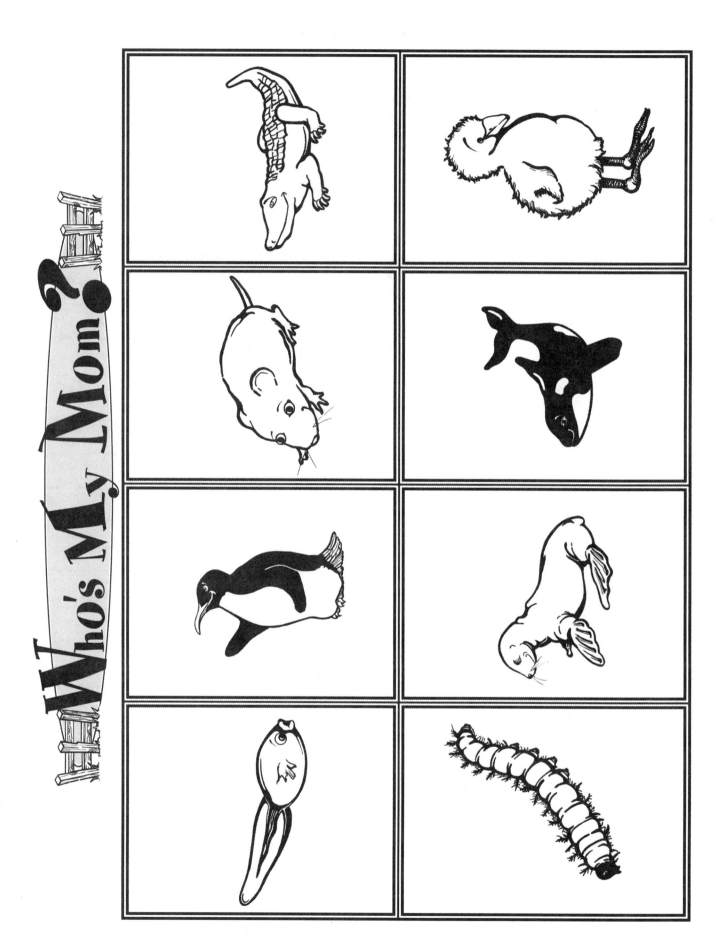

38

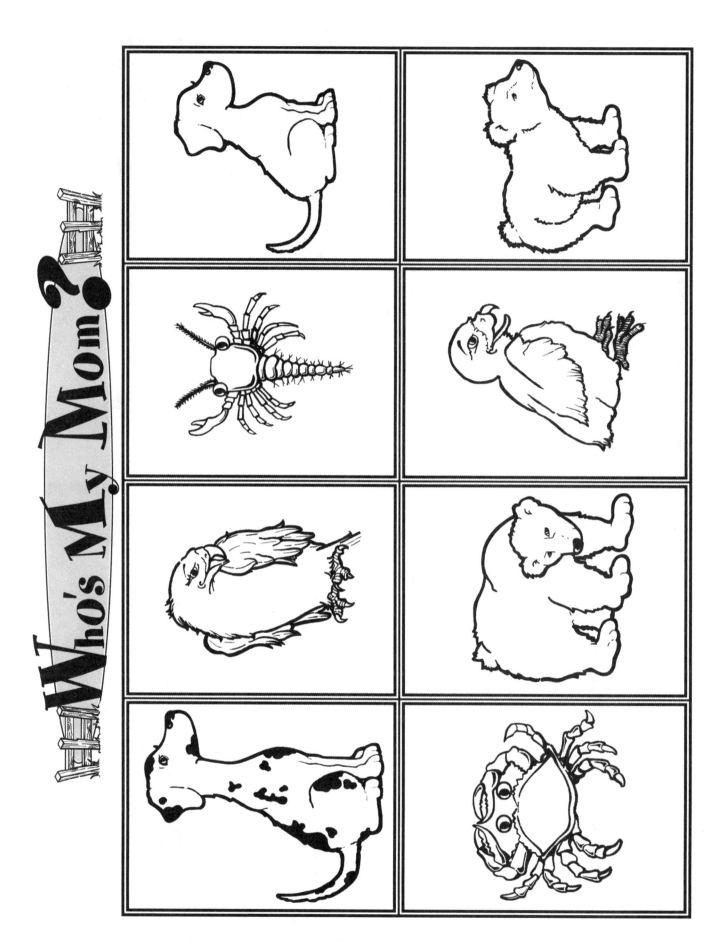

Meet the Guppy Family

Topic
Heredity

Key Question
How do baby guppies look like their parents?

Learning Goals
Students will:
- compare and contrast baby guppies with their parents, and
- learn to distinguish male and female guppies.

Guiding Documents
Project 2061 Benchmarks
- *Offspring are very much, but not exactly, like their parents and like one another.*
- *There is variation among individuals of one kind within a population.*

NRC Standards
- *Plants and animals closely resemble their parents.*
- *Many characteristics of organisms are inherited from the parents of the organisms, but other characteristics result from an individual's interactions with the environment. Inherited characteristics include the color of flowers and the number of limbs of animals. Other features, such as the ability to ride a bicycle, are learned through interactions with the environment and cannot be passed on to the next generation.*

Science
Life science
reproduction
heredity

Integrated Processes
Observing
Comparing and contrasting

Materials
For the class:
one male and one female guppy
one-gallon glass pickle jar or equivalent
aquarium gravel or sand
aquarium plants
guppy food
chart paper
fish net
water conditioner, optional

Background Information
The following information is intended to give the teacher enough information to feel comfortable leading the discussion with students over the several weeks or months over which the guppies are observed.

Guppies are probably the most popular of all aquarium fish. They are hardy, easy to care for, reproduce under ordinary conditions, and bear live young. Every birth event is cause for immediate observation and a class celebration.

Guppies reproduce sexually. Sperm from the male fertilizes an egg in the female.

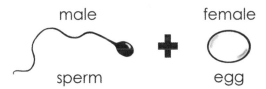

The abdomen on a pregnant female guppy has a dark spot called the *gravid spot*. The dark color is caused by the eyes of the baby guppies showing through the female's skin. The larger the abdomen and the darker the spot, the closer the female is to giving birth to the live-born baby guppies. Baby guppies are called fry. The term of the female guppy pregnancy is 22-26 days.

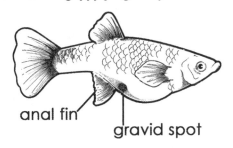

anal fin
gravid spot

The male guppy is the more colorful of the pair with a long, flowing tail. Typically, the male guppy is smaller than the female guppy. The anal fin of the male guppy is tube-shaped *(gonopodium)* to help guide sperm into the female during the mating ritual.

male guppy

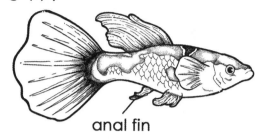

anal fin

The mating ritual begins with the male "dancing" around the female. The male's colorful tail can be seen vibrating rapidly, To complete the mating process, the male will use its anal fin (gonopodium) to place sperm into the female's eggs. Approximately three to four weeks after mating, anywhere from 10 to 50 young are born.

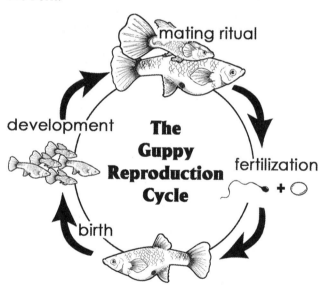

As the fry develop, observation makes it easy to identify which are males and which are females. After one month of development, the females are ready to breed.

In the wild, guppies are predators and feed on smaller fish and other aquatic organisms. They also eat mosquito larvae. Because the fry are so small, adult guppies will eat them and in this sense can be considered to be cannibalistic. In the classroom, isolate the parents from the fry by placing them in another aquarium.

No more than one pair of adult guppies should be put into a one-gallon container. A few aquarium plants like *Anacharis (Elodea)* should be put in the aquarium.

Guppies can survive in water that ranges from 65° to 85° Fahrenheit. If the temperature in the room is expected to get below 65° F, make arrangements to move the aquariums to a warmer location.

Guppies can be left uncared for over weekends and short school vacations. Over long vacations, make feeding arrangements or send a permission note to parents and distribute the fish to willing parents and students.

Management

1. Purchase a male and female guppy, aquarium gravel, fish nets, guppy food, and aquatic plants like *Elodea* (sometimes called *Anacharis)* at a local pet store that sells tropical fish.
2. Collect or purchase a one-gallon glass or clear plastic container. Delicatessens will often donate or sell empty glass pickle jars. Also, students may have suitable glass containers at home that parents would be willing to donate to the class.
3. Identify a place in the classroom, out of direct sunlight, where the aquarium can be stored when students are not observing it.
4. Do not overfeed the guppies, as overfeeding causes the water to sour. Feed the guppies a pinch of guppy food, ground between the thumb and forefinger, once in the morning and once in the late afternoon.
5. Over time, replace at least 50% of the water in the jar with fresh, aged water. Clean an empty gallon plastic milk container. Fill the container with tap water and let it sit for 24 to 36 hours. This allows any chlorine in the water to escape. The easiest method for replacing the water is to use a short length of plastic tubing to siphon the water from the jar into another container.

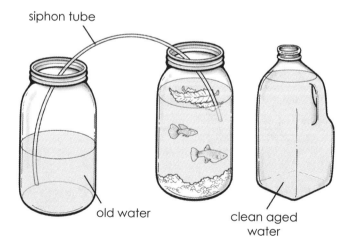

siphon tube

old water

clean aged water

After siphoning the water, put one hand in the jar and pour the fresh water from the gallon container onto your hand. This method minimizes the stress placed on the fish by changing the water.

Procedure

Introduction

1. Explain that the class is going to make a freshwater aquarium, stock it with aquatic plants and a male and female guppy.
2. Show the class the glass container. Cover the bottom of the container with aquarium sand or gravel to a depth of approximately one inch.
3. Fill the container with tap water to a level that maximizes the surface area of the water.

water level

4. Add the aquatic plants to the jar.
5. Set the jar at the spot you've designated for storing the aquarium. Tell students that the jar will sit for one day so that the water can reach room temperature and any chlorine in the water can evaporate.
6. The next day, gather the students together and add a male and female guppy to the jar.

Observing the Male and Female Guppies

1. Explain to the students that they are going to observe, over time, the male and female guppy in the jar, and notice if there are any changes in the guppies.
2. Ask the students how the guppies are similar and how they are different. Have them guess which guppy is the female.
3. Use the large picture of the female guppy to identify and describe the physical characteristics of the female guppy. [larger than the male, less colorful, has a dark spot at the rear of her stomach, the different fins, etc.]
4. Use the large picture of the male guppy to identify and describe the physical characteristics of the male guppy. [large, colorful tail; typically smaller than the female; spends a lot of time chasing (mating) the female; etc.]

5. Have students make daily observations of the guppies. Record these on chart paper.
   - Describe any changes in the physical characteristics of the female guppy.
   - Describe any changes in the physical characteristics of the male guppy.
   - Document how the gravid spot on the female guppy changes.
   - Document the birth event. Include date, time of day, behavior of both male and female, and the behavior of the fry.
   - Document the number of fry produced in each brood.

Observing the Development of the Fry

1. As soon as possible after the birth event, remove the male and female guppies from the jar.
2. Continue adding observations to the chart. Describe, over time, the development of the fry.
3. Document when it was possible to identify whether each fry was a male or female guppy.
4. Encourage students to describe the development of color patterns with the male guppies.
5. Compare and contrast the traits (size, color, etc.) of the developing young with the parents.

Connecting Learning

1. What does a female guppy look like?
2. How did she change over time? [development of the gravid spot]
3. What does a male guppy look like?
4. How did the male guppy change over time?
5. Describe the behavior of the fry after the birth event.
6. How did the fry change?
7. When was it possible to determine whether each is a female or a male?
8. How are baby guppies (fry) like their parents? How are you like your parents?
9. What are you wondering now?

Extension

Other tropical fish are also live-bearers. Repeat the activity with a pair of Mollys or Swordtails.

Female Guppy

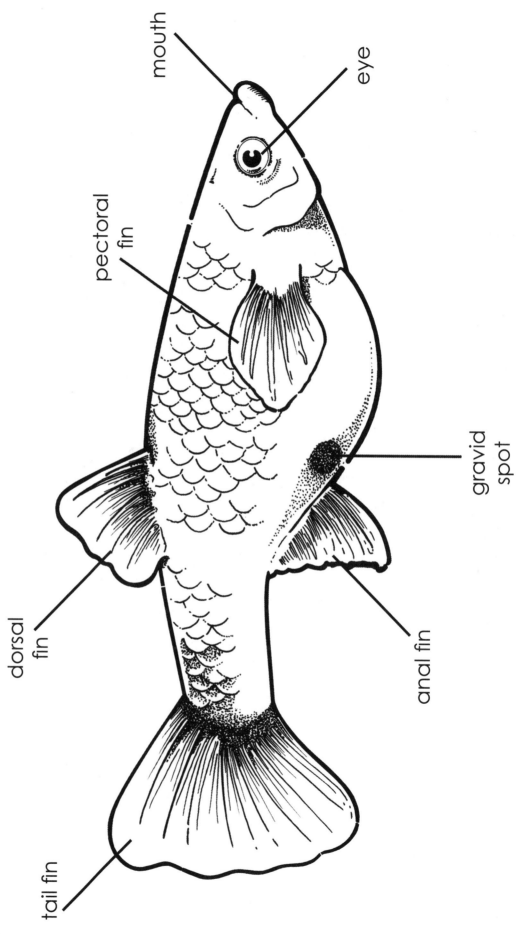

mouth

eye

pectoral fin

dorsal fin

gravid spot

anal fin

tail fin

Male Guppy

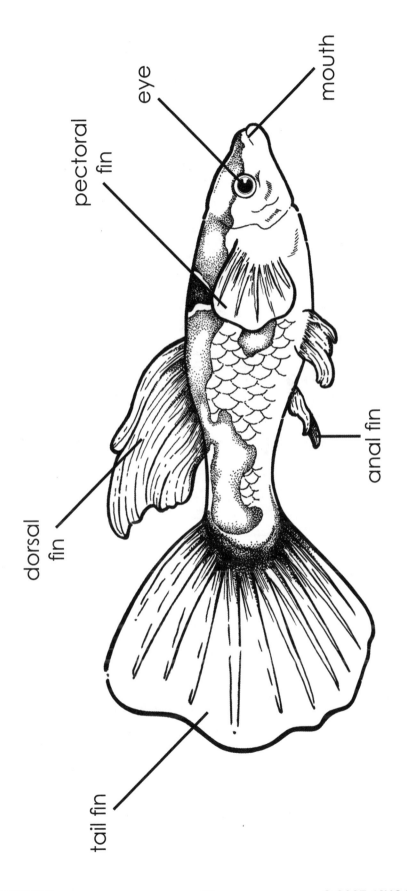

eye

mouth

pectoral fin

dorsal fin

anal fin

tail fin

Family Traits

Tune: Clementine

In a strol-ler sleeps a ba-by with his

bro-ther standing there. They will look like one a-

no-ther—light brown eyes and dark brown hair.

In a garden grows a daisy,
And its seeds begin to sprout.
They will grow up to be daisies.
We know this without a doubt.

In a pasture lives a field mouse,
And her babies have no fur.
In a few months they'll get bigger.
Soon they will look just like her.

We have learned that
Plants and creatures
That are all of the same kind
Tend to look like one another.
This is often what you'll find.

Wind Detectives

Topic
Wind

Key Question
What can we observe about wind and how it moves things?

Learning Goals
Students will:
- observe and describe evidence of wind,
- observe that objects are moved by the wind and that different objects move in different ways,
- draw pictures of things that are moved by the wind and sort them based on a variety of rules, and
- create a bar graph to display the pictures.

Guiding Documents
Project 2061 Benchmarks
- *People can often learn about things around them by just observing those things carefully, but sometimes they can learn more by doing something to the things and noting what happens.*
- *Things move in many different ways, such as straight, zigzag, round and round, back and forth, and fast and slow.*
- *Things move, or can be made to move, along straight, curved, circular, back-and-forth, and jagged paths.*

NCTM Standards 2000*
- *Sort and classify objects according to their attributes and organize data about the objects*
- *Represent data using concrete objects, pictures, and graphs*

Math
Data collection
  graphing
    bar graph
  sorting

Science
Earth science
  meteorology
    wind
Physical science
  force and motion

Integrated Processes
Observing
Comparing and contrasting
Collecting and recording data
Communicating

Materials
For Part One:
  Gilberto and the Wind (see *Curriculum Correlation*)
  various objects to observe in the wind
    (see *Management 3*)

For Part Two:
  crayons or colored pencils
  scissors
  glue
  butcher paper or chart paper (see *Management 4*)
  student pages
  magazines, optional

Background Information
The wind provides us with visible evidence of the invisible air around us. Wind is a result of the unequal heating of the Earth's surface that creates areas of high and low atmosheric pressure. Wind exerts a force on all objects it encounters. If this force is large enough to overcome the other forces acting on an object (friction and gravity), the object will move. This activity provides students with the opportunity to make some initial observations about the wind and how it moves different objects in different ways. Students should be encouraged to observe the changes in the wind day by day and over time. Emphasis should be placed on students making careful observations.

Management
1. This activity should be done on a windy or breezy day.
2. This activity works best as a total group activity, followed by small group and individual explorations.
3. Prior to doing this activity, gather a collection of objects that will be moved easily by the wind such as pieces of paper, Styrofoam cups, plastic bags, leaves, feathers, tissue, and so on. Also collect a few items that will not move easily in the wind (rocks, staplers, textbooks, etc.) and a few that may move depending on how they are positioned (round pencils, small paperback books, etc.),

4. You will need a piece of chart paper or butcher paper large enough to use as a whole-class bar graph on which to organize and display students' drawings from *Part Two*.

Procedure

Part One

1. Have students discuss what they know about wind. Encourage the students to look out the window and decide if the wind is blowing. Have them discuss the evidence that the wind is blowing.

2. Read the story *Gilberto and the Wind* (see *Curriculum Correlation*) and ask the students if they have ever played in the wind. Ask how they can tell that the wind is blowing in the story and what things it moves.

3. Brainstorm some objects that the wind can move and record them on a chart or board.

4. Take the students on a walk outside to observe evidence of the wind on the playground. [moving trees, blowing swings, flag flapping, dust devils, etc.] Discuss the different ways in which the wind can be felt and seen.

5. Return to the classroom and distribute to the students as many of the items from the list as possible. Supplement the items brainstormed with other items suggested in *Management 3* so that each child has one item. At least two or three children should have objects that will not move as easily in the wind, and two or three children should have objects that may or may not move depending on how they are placed (e.g., round pencils, paperback books, empty margarine tubs, etc.).

6. Go to a blacktop or other relatively smooth surface that is exposed to the wind. Have each student place his or her object on the ground so that the class can observe how the wind affects it. (Be sure to have students pick up any objects that do not belong in the natural environment once they have been observed.)

7. Have students with objects that were not moved by the wind experiment with different ways to place the objects that might cause them to be moved. For example, a round pencil would not roll when placed parallel to the wind, but might roll when placed perpendicular to the wind.

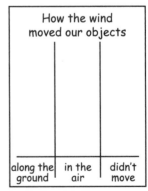

8. Help students verbalize that the wind moves some things more easily than others. Help them recognize that even the way an object is placed can change whether or not it will be moved by the wind. Have them use descriptive language to compare and contrast the different ways in which the different objects moved.

Part Two

1. Take the students back inside and distribute the first student page and the crayons or colored pencils.

2. Direct the students to draw a picture in each box to record something the wind can move. (One of the four pictures should be of something that was not observed in *Part One*.) For younger students, pictures can be cut from magazines and pasted into the boxes.

3. When students have finished, direct them to cut their papers into the four sections. Have them work in small groups to sort the pictures into two or more groups. Rules for the sorts can be given by you as the teacher, or come from students, but they should all have to do with the objects in relation to the wind and how they moved. Examples of sorts could include: things that the wind moved all the time, things that the wind moved some of the time, and things that the wind never moved; things that moved quickly in the wind and things that moved slowly in the wind; things that moved along the ground and things that moved in the air, etc.

4. When the groups have sorted their pictures, invite them to share their sorting strategies. Continue to share ideas and encourage as many sorting options as possible.

5. As a class, decide on one sorting scheme to use for all the pictures. Make a bar graph on a sheet of chart paper or butcher paper in which to display the pictures. Give the graph an appropriate title and label each section.

How the wind moved our objects		
along the ground	in the air	didn't move

6. Place the graph on a low table or on the floor. Have the groups come up one at a time and place their drawings in the appropriate sections on the graph. Once all groups have sorted their drawings, have the class gather around the graph and decide if they all agree with the placement of the pictures.

48

If there are any disagreements, have the students who placed the pictures explain why they made the choices they did.

7. When there is consensus about the organization of the graph, glue down each picture so that the graph can be displayed on the wall.

8. Distribute the second student page and have students either write or draw in the space provided at least two things that the wind can do. Encourage students to draw upon their experiences from the activity when completing this page.

Connecting Learning

Part One

1. What evidence of the wind did we see on the playground? [tree branches moving, hair blowing, flags waving, dust in the air, etc.]

2. What were some of the objects that moved in the wind?

3. What were some of the objects that did not move in the wind? Why didn't they move? [too heavy, wind wasn't strong enough, etc.]

4. What were some objects that only moved if they were placed a certain way? [pencil, cup, etc.] How did they have to be placed so that they would move? [facing the wind, on their sides, etc.]

5. How were the objects that moved in the wind different from those that did not move? [lighter, less stable, smaller, etc.]

6. Describe the different ways our objects moved in the wind. [some moved along the ground, some blew up into the air, some flapped or fluttered without moving very much, some moved quickly, some moved slowly, etc.]

7. What could you do to keep the wind from moving an object? When would you want/need to do that?

8. Why would it be important to know if the wind is blowing? Who are the people who need to know?

Part Two

1. What pictures did you choose to draw? Why?

2. What were some of the ways your group sorted your pictures? Why did you choose those sorts?

3. What does our graph tell us about the objects that the wind can move?

4. What are some of the things the wind can do?

Extensions

1. Have students take some of the objects from *Part One* outside on a windy day and observe what they do when they are not placed on the ground. How many of them still move when they are held in the hand? Give students ribbons, crepe paper, or similar objects that will flutter when held and have them compare and contrast the characteristics between the two sets of objects.

2. Have students brainstorm and experiment with objects that may or may not move in the wind depending on the state they are in. For example, a closed umbrella would probably not move in the wind, while an open umbrella likely would. Other objects could include inflatable balls, balloons, or cardboard boxes (broken down vs. assembled).

3. Help the students make their own wind scale. For example:

   No Wind—the flag hangs limply, the trees are still

   Light wind—the wind felt on face, the leaves are barely moving

   Moderate wind—light flag extended, small branches move, pieces of paper and trash are blown around

   Strong wind—large branches move, flag is flapping vigorously

   Dangerous wind—difficult to walk against wind, small branches broken off trees

Curriculum Correlation

Literature

Ets, Marie Hall. *Gilberto and the Wind*. Viking Press. New York. 1963.

Simon, Seymour. *Let's Try It Out in the Air*. Simon & Schuster. New York. 2001.

* Reprinted with permission from *Principles and Standards for School Mathematics*, 2000 by the National Council of Teachers of Mathematics. All rights reserved.

Wind Detectives

Wind Detectives

The wind can...

Blow and Go

Topic
Force and motion

Key Question
How many times do you have to blow an object to make it travel one meter?

Learning Goals
Students will:
• use moving air to push objects, and
• collect and record data about their experiences.

Guiding Documents
Project 2061 Benchmarks
• *Things move, or can be made to move, along straight, curved, circular, back-and-forth, and jagged paths.*
• *The way to change how something is moving is to give it a push or a pull.*
• *Numbers can be used to count things, place them in order, or name them.*
• *Simple graphs can help to tell about observations.*

NRC Standards
• *The position of an object can be described by locating it relative to another object or the background.*
• *The position and motion of objects can be changed by pushing or pulling. The size of the change is related to the strength of the push or pull.*

NCTM Standards 2000*
• *Pose questions and gather data about themselves and their surroundings*
• *Represent data using concrete objects, pictures, and graphs*

Math
Estimation
Counting
Graphing

Science
Physical science
    force and motion
    wind energy

Integrated Processes
Observing
Comparing and contrasting
Collecting and recording data
Interpreting data

Materials
For each group:
    four classroom objects (see *Management 3*)
    crayons
    student pages
    scissors, optional
    glue sticks, optional

For the class:
    masking tape
    meter stick

Background Information
In simple terms, forces are pushes and pulls that can make things move. Some forces are contact forces; they require direct contact (touching) to make the object move. For example, you apply a force when your hand pushes a door to open it. When your hand pulls on a hose, you apply a force to move it. When you blow on a piece of paper to move it, your breath comes in contact with the paper. There are also non-contact forces. These are forces that act from a distance without making direct contact with the object. Gravity and magnetism are two non-contact forces.

Management
1. This activity works best in small groups.
2. For each group, place a piece of masking tape about one meter long on the floor for the starting line. Place another piece of masking tape for the finish line one meter from the start.

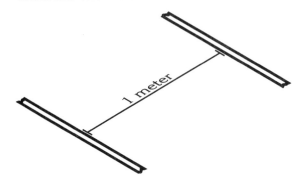

1 meter

52

3. Collect various classroom/household objects to use in the investigation. Each group should have four objects. It may be helpful if two of the objects are the same for all groups. Some objects you might use are: marbles, paper, clothespins, pencils, rubber balls, facial tissues, crayons, leaves, paper clips, etc.

4. Have students use crayons to record their estimates and the actual results. It is a temptation to erase in order to be "right," and crayons may help them see that an estimate may not always be "right," but the goal is to improve the estimates by having more chances to practice and make observations.

Procedure

1. Discuss what the students think they can do to move objects from one point to another.

2. Ask students what things they think they can move just by blowing on them.

3. Ask the students if they think they can move a pencil from the starting line marked on the floor to the finish line marked by blowing.

4. Ask one student to demonstrate blowing the pencil from the start to the finish line. Discuss the number of times the student needed to blow on the pencil.

5. Give each group their set of four objects. Distribute a record page to each student. If you are using objects pictured on the included page, distribute that as well.

6. Have students fill in the objects they will be using on their graphs. They can draw a picture, write the word, or cut out and glue a provided picture. Explain that they will need to estimate the number of times they will have to blow on each object to move it from the start to the finish line. Have them record their estimate for the first object only in the "I think" column.

7. Have groups go to their areas and take turns testing the first object, with each student recording the actual number of blows needed on his or her own record page.

8. Repeat this process with the remaining objects (estimate, blow the object, record the actual results).

9. When all groups have had a chance to finish, gather together in a group area and discuss the results of their investigation.

10. Students may want to post the record sheets and discuss the results that other groups got.

Connecting Learning

1. What caused the objects to move?

2. Were you able to move all of the objects when you blew on them? Explain.

3. Did all of the objects move with the same number of blows? Explain.

4. Which object took the fewest blows? Which took the most? Why?

5. Were there some objects that wouldn't move at all? Why do you think this happened?

6. How did your estimates compare with your actual results? Did it get easier to estimate as you went along? Why or why not?

7. If you could conduct the investigation again, what would you change? Why?

8. If you could choose four new objects, what would you choose? Why would you choose them?

9. Did it make a difference who was blowing on the objects? Why?

10. What are you wondering now?

Extensions

1. Repeat the activity using a straw to direct the air.

2. Choose objects that have a shared attribute (all cylinders, spheres, etc.) and see how they compare.

3. Read *The Three Little Pigs* and discuss how the wolf was able to blow down the pigs' houses. Build some houses and see if the children are able to blow them down.

4. Read *The True Story of the 3 Little Pigs!* by Jon Scieszka and talk about the similarities and differences between this and *The Three Little Pigs*.

Curriculum Correlation

Scieszka, Jon. *The True Story of the 3 Little Pigs!* Puffin Books. New York. 1966.

Home Link

Send a piece of adding machine tape, one meter long, home with each student. Tell students to work with their parents to set up a course at home and try four objects. Ask them to keep a record of what happened and be ready to share the results with the class.

* Reprinted with permission from *Principles and Standards for School Mathematics*, 2000 by the National Council of Teachers of Mathematics. All rights reserved.

Make a guess. Record the results.

Number of Blows	I think	I found out	I think	I found out	I think	I found out	I think	I found out
10								
9								
8								
7								
6								
5								
4								
3								
2								
1								

Objects

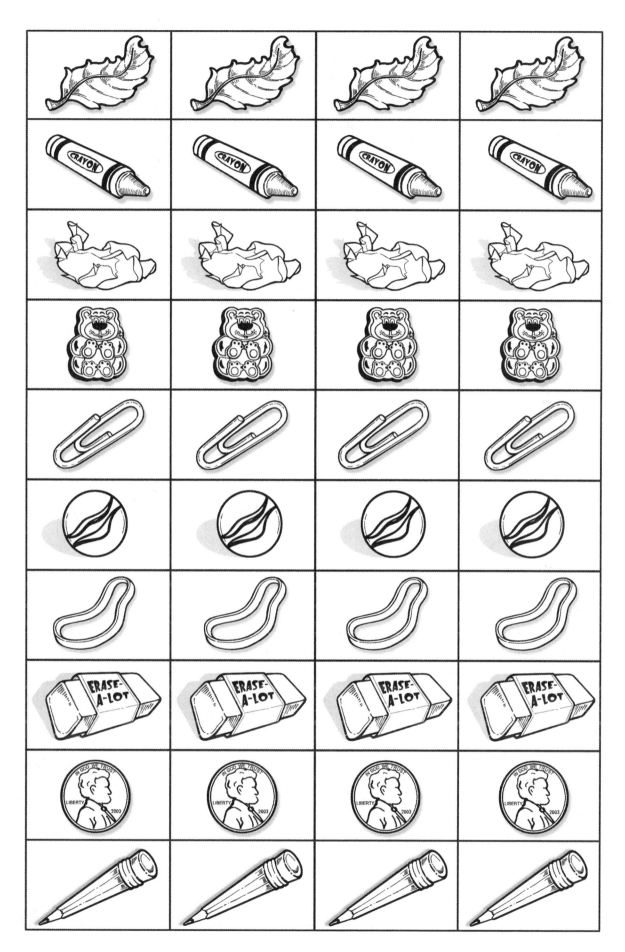

Feather Relays

Topic
Force and motion

Key Question
How can we use our own breath (air) to move a feather in a relay?

Learning Goals
Students will:
- use their breath as a force to move feathers;
- observe, compare, and communicate their observations; and
- participate in a feather relay.

Guiding Documents
Project 2061 Benchmarks
- *Things near the earth fall to the ground unless something holds them up.*
- *Things move in many different ways, such as straight, zigzag, round and round, back and forth, and fast and slow.*
- *The way to change how something is moving is to give it a push or a pull.*
- *Describe and compare things in terms of number, shape, texture, size, weight, color and motion.*
- *Raise questions about the world around them and be willing to seek answers to some of them by making careful observations and trying things out.*

NRC Standard
- *The position and motion of objects can be changed by pushing or pulling. The size of the change is related to the strength of the push or pull.*

Science
Physical science
  force and motion
    pushes

Integrated Processes
Observing
Comparing and contrasting
Collecting data
Applying

Materials
For each student:
  feather or other light object (see *Management 1*)
  balloon (see *Assessment*)
  My Book About Air page

For each group of students:
  relay course (see *Management 3*)

For the class:
  a version of *The Three Little Pigs* Story or *The True Story of the 3 Little Pigs!* (see *Curriculum Correlation*)

Background Information
Young children are naturally curious about themselves and the world around them. They ask questions about and have experience with many moving things—themselves, toys, leaves, water, etc. It is important that they have numerous opportunities to observe, compare, and talk about the way objects move and the way they can move things by pushing or pulling. As they observe and compare, they build the foundation for future understandings through simple investigations.

Management
1. Have students bring in a feather to use or purchase some at a craft store.
2. Arrange students in groups of four or five with an adult volunteer or cross-age tutor. Have several groups working at the same time or use as a center activity.
3. A table makes a good relay course. One end is the starting line, the other end is the finish line. Children can stand or kneel around the table, depending on the table's height. If a table is unavailable, mark the course on the floor using masking tape for a start and a finish line and have children positioned along the course. The course should be 2-3 meters (6-10 ft) in length.

Procedure

1. Discuss with students their understanding of air. Ask, "What things can you do with your own air?"
2. Read a version of *The Three Little Pigs*, and/or use the book *The True Story of the 3 Little Pigs!* with the children in order to stimulate their thinking about puffs making things move.
3. Discuss the different things that the wolf blew down in either story.
4. Discuss some of the things they think they could move by blowing them.
5. Share the feathers the students have brought in or the feathers purchased from a craft store. Have the students compare the similarities and differences among the feathers.
6. Give each student a feather and have them take a few minutes to explore moving the feather in various ways and share their observations.
7. Tell the students that they will work in small groups and conduct a feather relay.
8. Show them the relay course and explain how the race will be done. Each member of the team will be equally spaced along the relay route. The first person will blow the feather until it reaches the second person. Then the second person will then take over and blow the feather to the third person, and so on. The final person must blow the feather over the finish line.
9. Instruct each group to observe and be ready to share the results of their relay. Establish a starting and ending time for the activity and send groups to begin the relay.
10. After all students have had time to participate in the feather relay, bring them together and share their observations.
11. Distribute the *My Book About Air* page and show students how to fold it in half horizontally and then vertically to make a book.
12. Instruct them to draw or write about what happened during the feather relay. Then have them illustrate or write about two other things they can move with their air and two things they cannot move with their air.
13. Collect the books and save for the assessment with balloons.

Connecting Learning

1. What happened when you blew your feather? What made your feather go the way you wanted it to? Were there any strategies you used to help you blow the feather in the relay? Explain.
2. When your group tried different feathers or saw other groups' feathers, were the results the same? Explain. How were the feathers the same? ...different? Why do you think the feathers behaved the way they did?

3. How did puffing move your feather? [The air from my mouth pushed the feather.]
4. Besides blowing, how else could you make your feather move? [fan it, use an electric fan, blow through a tube or straw, etc.]
5. What other questions do you have about what you can puff? How could you find the answers to your questions?

Assessment

Students need the opportunity for multiple investigations before they begin to understand a concept. After using a feather, set up a similar investigation using a balloon as an assessment. Observe students as they repeat the relay to see if they are using their observations from the feather race when they blow on the balloon. Listen to their conversations to hear generalizations or applications of previously learned concepts. Are they talking about objects in motion? Are they talking about air as the push for moving the balloon? Are they asking questions about other objects that they could make move? Give the students their *My Book About Air* and have them illustrate or write about what happened during their balloon relay. Ask students how the balloon and feather relays were alike and how they were different.

Extensions

1. Repeat the investigation using other lightweight objects such as tissue, crepe paper, etc.
2. Repeat the investigation using a different wind source: fan, bellows, hair dryer, etc.

Curriculum Correlation

Sciezka, Jon. *The True Story of the 3 Little Pigs!* Penguin Books. New York. 1999.

Home Link

Have students find some things at home that they can make move with their "puff." Direct them to draw a picture of each object they were able to move and to be ready to share their drawings the next day.

3

2.

1.

Two things I cannot
move using my air:

2.

1.

Two things I can move
using my air:

This is what happened
during the feather race.

This is what happened
during the balloon race.

My
Book
About

Air

4

1

Wind in Your Sails

Topic
Wind energy

Key Question
Which of the sail shapes makes the fastest boats?

Learning Goals
Students will:
- compare the speeds of boats that have different-shaped sails, and
- determine which sail shape makes the fastest boats.

Guiding Documents
Project 2061 Benchmarks
- *Circles, squares, triangles, and other shapes can be found in things in nature and in things that people build.*
- *Shapes such as circles, squares, and triangles can be used to describe many things that can be seen.*
- *Things move, or can be made to move, along straight, curved, circular, back-and-forth, and jagged paths.*
- *Things move in many different ways, such as straight, zigzag, round and round, back and forth, and fast and slow.*
- *The way to change how something is moving is to give it a push or a pull.*
- *A model of something is different from the real thing but can be used to learn something about the real thing.*

NRC Standards
- *The position and motion of objects can be changed by pushing or pulling. The size of the change is related to the strength of the push or pull.*
- *Plan and conduct a simple investigation.*

NCTM Standards 2000*
- *Pose questions and gather data about themselves and their surroundings*
- *Recognize geometric shapes and structures in the environment and specify their location*

Math
Geometry
  2-D shapes

Science
Physical science
  wind energy

Integrated Processes
Observing
Comparing and contrasting
Communicating
Generalizing

Materials
For the class:
  rain gutter for "rivers" (see *Management 5*)
  fan (see *Management 6*)
  prediction graph (see *Management 7*)

For each student:
  empty half-pint milk carton (see *Management 2*)
  one sail pattern (see *Management 3*)
  coffee stirrer
  scissors
  transparent tape
  clay
  sailboat marker

Background Information
People have always known the power of wind, and have been using it to perform a variety of tasks for many thousands of years. Sailboats are a common and enduring example of using wind energy to generate motion. Sailors are experts at manipulating sails and their boats to make the most of any wind that's available. Sails come in a variety of shapes and sizes depending on the kind of boat and purpose of the sail.

In this activity, students will create simple boats from half-pint milk cartons and give them sails cut from pieces of card stock. They will then perform a series of races between two boats at a time. Through these direct comparisons, they will determine which sail shape makes for the fastest boats.

Management
1. This activity is divided into two parts. They can both be done on the same day, or spread out over two days.
2. Collect at least one half-pint milk carton per student. Rinse the cartons and allow them to dry. Tape the open end shut and cut off one side. If milk cartons are not available, small butter tubs or other identical plastic containers may be used instead.
3. Copy the page of sail patterns onto card stock. Make enough copies so that each student will have one sail. Cut the pages into quarters before distributing the patterns to students.
4. Make enough copies of the sailboat markers so that each student can have one.
5. You will need to set up two "rivers" side by side so that pairs of students can race their boats. Purchase two

59

lengths of plastic rain gutter and cap the ends for a watertight container. Fill the gutters half-full with water.

6. You will need a fan that has a low speed setting to provide the wind energy for the sailboats. Before doing the activity with students, experiment with the proper distance to place the fan from the rain gutters so that it will gently move boats in both gutters.

7. Create a prediction graph on a piece of chart paper. At the top write, "Which sail shape will make the fastest boats?" Make five columns—one for each of the shapes [square, rectangle, equilateral triangle (triangle one), right triangle (triangle two)], and one labeled "shape won't matter."

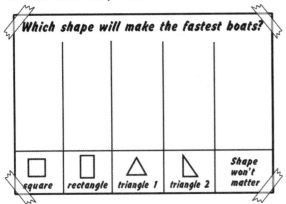

Procedure

Part One—Creating the Boats

1. Ask if anyone has ever been on a sailboat. Invite a few students to share about the experience. Ask them to describe what they can remember about the sails—their size, shape, and how they worked.

2. Have students identify the purpose of the sails. [to catch the wind and cause the boat to move] Discuss how, before engines were invented, all boats had to use the wind or manpower (rowing) to move.

3. Tell students that they will be building boats and experimenting with different kinds of sails. Distribute the sail patterns and have students cut them out.

4. Have students identify the different shapes they have for their sails. [triangle, square, rectangle]

5. Give each student a coffee stirrer. Instruct them to tape the stirrers in the spaces indicated on the sails.

6. Distribute the milk cartons and clay. Show students how to roll the clay into a ball, stick it to the inside bottom of the milk carton, and insert the coffee stirrer into the clay. Assist students as necessary so that the sails are all in relatively the same position within the boats.

7. Tell students that they will be racing their boats in the two rivers you have set up, but that

first, you would like them to make some guesses about how the different sails will affect the speed of the boats.

8. Show the class the prediction chart. Distribute a sailboat marker to each student. Have them write their initials on the markers and tape them to the prediction graph in the columns that reflect which sail shape they think will be the fastest.

Part Two—Racing the Boats

1. Invite the class to gather around the rivers. Select two students with the same sail shape and invite them to place their boats in the rivers. Turn on the fan and have the students hold their boats in place at the ends closest to the fan.

2. At the count of three, have students release their boats and step back to observe what happens. The first boat to reach the opposite end of the river is the winner.

3. Repeat this race, pitting the winner against another boat with the same sail shape until the fastest boat with one sail shape is determined. (If desired, you can race each pairing multiple times and take the winner as the best two out of three races.)

4. Repeat for all the other sail shapes until the fastest boat with each sail shape is determined.

5. Race the four fastest boats against each other, two at a time, until the fastest boat overall is determined.

6. Discuss what students observed and how the results compared to the predictions.

Connecting Learning

1. What made our boats move? [the wind from the fan pushed on the sails] How do you know?

2. How are our boats like real sailboats? How are they different?

3. Which sail shape did you predict would be the fastest? Why?

4. Which sail shape turned out to be the fastest?

5. Do you think the way we found the winner was fair? Why or why not?

6. Do you think your results would have been different if you used cloth or plastic instead of paper? Why or why not?

7. If your boat were made in a different shape, do you think the same sail shape would be best? Explain.

8. What are you wondering now?

Extensions

1. Repeat the activity using different materials for the sails and/or different boat designs.

2. To discover the effects of multiple sails, continue testing using more than one sail.

Wind in Your Sails
Sail Patterns

Copy this page onto card stock. Make enough copies so that each student will have one sail pattern. Cut the page along the dashed lines before giving the sails to students to cut out.

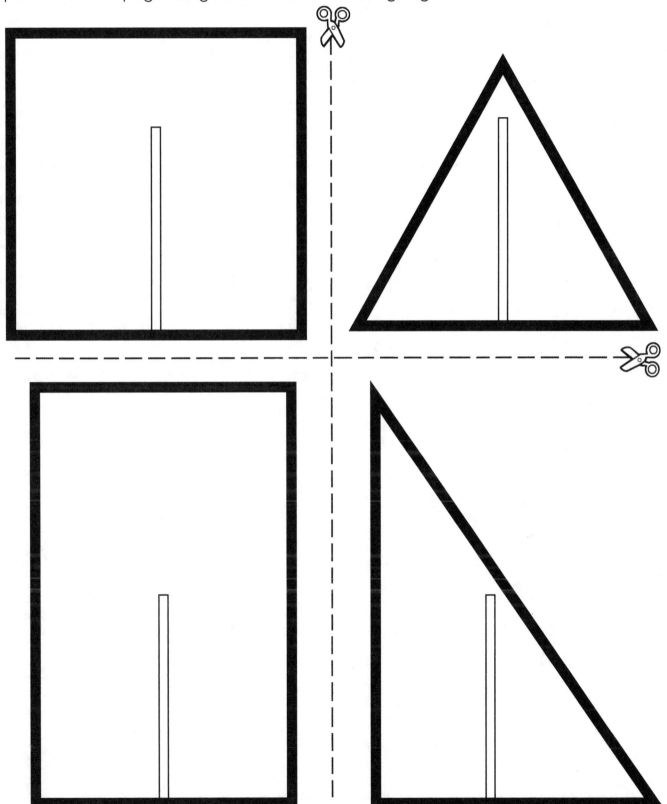

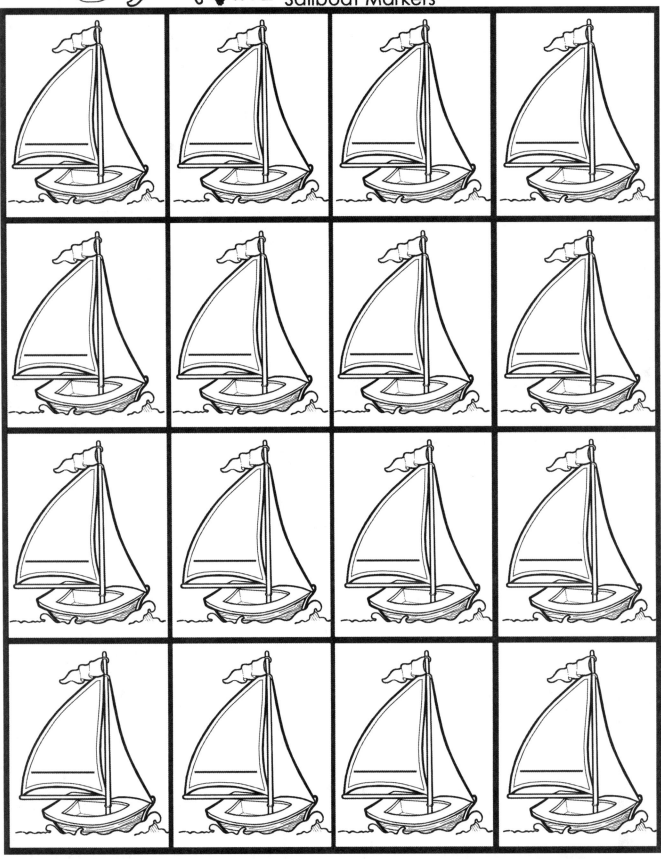

Wind in Your Sails
Sailboat Markers

Topic
Force and motion

Key Question
How are the concepts of push and pull important when flying a kite?

Learning Goals
Students will:
- define and explore the concepts of push and pull,
- relate push and pull to real-world situations such as flying a kite,
- construct simple kites, and
- fly the kites to experience how push and pull are important in keeping the kites in the air.

Guiding Documents
Project 2061 Benchmarks
- *The way to change how something is moving is to give it a push or a pull.*
- *Things near the earth fall to the ground unless something holds them up.*

NRC Standard
- *The position and motion of objects can be changed by pushing or pulling. The size of the change is related to the strength of the push or pull.*

NCTM Standards 2000*
- *Understand how to measure using nonstandard and standard units*
- *Measure with multiple copies of units of the same size, such as paper clips laid end to end*

Math
Measurement
    length

Science
Physical science
    force and motion
        push and pull

Integrated Processes
Observing
Comparing and contrasting
Analyzing
Generalizing

Materials
For each student:
    one 12" x 18" sheet of construction paper
        (see *Management 4*)
    10-15 meters of kite string (see *Management 11*)
    one non-flex drinking straw (see *Management 10*)

For the class:
    paper, various sizes, colors, and types
        (see *Management 4*)
    crayons or colored pencils
    cotton balls (see *Management 6*)
    scissors
    tape
    glue
    lion and lamb patterns, enlarged
        (see *Management 5*)
    Kite Flying Safety sheet

Background Information
National standards at the primary level talk about force and motion in terms of pushes and pulls. Force is different from other measurable quantities such as mass or length because it has a direction attached to it. You cannot exert a force with no direction. Even the youngest children intuitively understand this concept. They know that if they push on something, it will move away from them (if the magnitude of the push is great enough), and that if they pull something, it will move toward them (if the magnitude of the pull is great enough). They understand that you cannot push or pull something without moving it unless the object is too large or heavy to be moved by the strength of the push or pull.

In this activity, students will be exploring pushes and pulls in the context of flying a kite. The push, in this case, is exerted by the wind on the surface of the kite, causing it to fly in the air. The student holding the string attached to the kite exerts the pull. As long as the magnitude of the push and the pull are constant, the kite will fly in the same position. If the wind suddenly begins to blow harder, the push will cause the student to feel a stronger tug on the line. This will force them to pull against the kite with more strength or move forward. Likewise, if the student pulls on the string causing the strength of the pull to be greater than the push of the wind, the kite will move toward the ground and possibly dip, or even fall from the sky.

63

Management

1. This activity is divided into two parts. In *Part One*, students will be challenged to think about force and motion as pushes and pulls. In *Part Two*, students will apply the concepts of push and pull to the real world as they fly kites and experience the push of the wind and their pull on the string.

2. You will need a windy day to do *Part Two* of this activity. The one-straw kites that students will be making require a consistent breeze in order for them to experience the feel of the wind pushing on the kite.

3. Please adhere to the safety rules when flying kits.

4. To make the bodies (heads) of the kites, students will each need one 12" x 18" piece of construction paper. White or cream is recommended for the lamb, and yellow for the lion. To decorate the kites and to make tails, tissue paper, crepe paper, and/or construction paper in a variety of colors and sizes should also be made available.

5. There are two kite patterns from which students can choose—a lion and a lamb. You may wish to let students choose their patterns, or simply give each pattern to half of the class. To make the kites, enlarge the patterns 125% so that they will be the right size for a 12" x 18" sheet of paper. A sufficient number of patterns to be used as templates should be copied onto cardstock and cut out before beginning this activity.

6. Students should be encouraged to decorate their kites using the materials provided. To make a mane for the lion, students can cut a fringe out of brown or yellow tissue or crepe paper and glue it to the perimeter of the kite. (Construction paper is not as desirable because of its weight.) To decorate the lambs, students can glue small amounts of cotton from separated cotton balls to the face and tail. Be cautious about adding too much weight as the weight of the kites will affect the amount of wind necessary to get them to fly.

7. You will have to decide if you want to have the kites cut out for your students ahead of time, or if you want them to do the tracing and cutting themselves.

8. A step-by-step construction guide for the kites is included following the teacher's manual. Be sure to make at least one kite yourself before doing this activity with the class to be sure that you understand the procedure.

9. The 3-cm fold that is made to create the keel of the kite is very important. Unless your students are able to fold neatly and accurately, you will need to make this fold for them once the patterns are cut out.

10. Do not use flexible straws.

11. Each student should be given 10-15 meters of kite string. This length will allow sufficient height but will be short enough for students to deal with. The string should be wound around a piece of cardboard, an empty paper tube, or some other similar object. Each piece of string should be securely taped to the holder so that it does not pull free while students are flying the kites. Another option is to have students bring kite string from home.

Procedure

Part One

1. Ask the class what a push looks like. Invite several students to come up to the front of the class and demonstrate a push. Have each one do something different.

2. Ask them what all of the pushes had in common. (Most of the students probably used their hands to put pressure against something that moved away from them as a result of the force.)

3. Ask students if they think it is possible to push something without touching it. Brainstorm possibilities as a class. Hopefully someone will eventually think of using his or her breath (wind) to push something. (If not, make the suggestion yourself.)

4. Challenge students to push an object, such as a pencil, across their desks using their breath.

5. Discuss whether students think a push using their hands and a push using their breath are the same. Be sure that students understand that even though they can't see their breath, it pushes on objects in exactly the same way that their fingers do.

6. Have students think of examples in the real world where things are pushed by air. [Trash blowing in the wind, kites, dandelion seeds, etc.]

7. Ask the class what a pull looks like. Invite several students to come up to the front of the class and demonstrate a pull. Have each one do something different.

8. Ask them what the pulls had in common. (Most of the students probably used their hands to put pressure on something that moved towards them as a result of the force.)

9. Have students think of examples in the real world where things are pulled. [Trailers attached to cars, pulling a wagon down the street, being pulled behind a boat while water-skiing, etc.]

10. Discuss the difference between a push and a pull. [Both pushes and pulls are forces exerted on objects. The difference is the direction of the force.]

11. Focus the discussion on kites. Ask students what makes a kite fly. [the wind] Ask what keeps a kite from flying away when it is being pushed by the wind. [It is attached to a string that someone is holding.]

Part Two

1. Tell students that they are going to be making their own kites so that they can explore pushes and pulls.

2. Distribute the necessary materials for the kites and assist students in cutting them out and taping them together (if this has not been done for them).

3. Once students have cut out the faces of their kites, distribute the materials for decorating and have students make their lions and lambs. Be sure that they decorate the side of the kite that will be facing the ground when it flies. This is the side with the keel.

4. Assist students in taping the straws to the backs of the kites and affixing the strings to the proper places on the keels.

5. Have students measure the heights of their kites using a non-standard unit such as paper clips in a chain, Unifix cubes, a piece of string, etc. Instruct them to design tails that are about three times as long as their kites are high. Tails need not be extremely wide (2-5 cm). Assist students in taping the tails to the bottoms of their kites. (Keep in mind that the larger the tail, the greater the drag (air friction). If a student's kite does not fly, it may have too much drag. Shortening or perhaps narrowing the tail can solve this.)

6. Go over the safety rules page with students and establish the rules for flying the kites in this activity. Be sure to set a distance that students must keep from each other while flying their kites.

7. Take the class outdoors to the open space you plan to use. Discuss how to get the kites to fly.

8. Assist students in launching their kites and try to help them keep the kites at a good "cruising altitude." (Depending on the wind conditions and the design of the kites, this may take some time and patience to achieve.)

9. If any students have kites that have trouble flying, try to make the necessary modifications so that they will be able to fly them. (See *A Word About Kites*.)

10. Ask students to pay attention to how it feels to hold the kite while it is flying.

11. At various times, instruct students to pull on their kite strings and observe what happens.

12. Instruct students to change their positions in relation to the direction of the wind and to note the behavior of the kites.

13. After students have had sufficient time to fly and explore, have them slowly and carefully reel in their kites and come inside.

14. Once back in the classroom, have a closing time of discussion and sharing where students share their observations and what they learned from the experience.

Connecting Learning

1. How did you have to stand in order for your kite to fly? [with my back to the wind] Why? [so that the wind can push on the front surface of the kite]

2. What happened when you moved so that the wind was not at your back? [The kite would not fly.] Why?

3. How did it feel to hold the kite while it was flying? [The wind could be felt pushing against the kite, the string was straight and taut, etc.]

4. What happened when you pulled the kite string? Did how hard you pulled change how the kite reacted? [The harder the pull, the more the kite is affected.]

Curriculum Correlation

Literature

Cobb, Vicki. *I Face the Wind*. HarperCollins. New York. 2003.

Ets, Marie Hall. *Gilberto and the Wind*. Puffin Books. New York. 1963.

Murphy, Stuart J. *Let's Fly a Kite (MathStart)*. Harper-Collins. New York. 2000.

A Word About Kites

Even the best of store-bought kites can be temperamental and difficult to fly. Homemade kites have even more potential to be poor fliers. Many factors can influence how well a homemade kite will fly. Here are some suggestions for ways to fix common problems.

If the kite:	Try:
won't fly at all.	reducing the amount of drag by shortening the tail or moving the string up on the keel.
flies, but then circles and crashes.	moving the string down on the keel.
flies, but is unstable.	lengthening the tail to increase the drag.

* Reprinted with permission from *Principles and Standards for School Mathematics*, 2000 by the National Council of Teachers of Mathematics. All rights reserved.

Flying Lion, Gliding Lamb

Kite Flying Safety

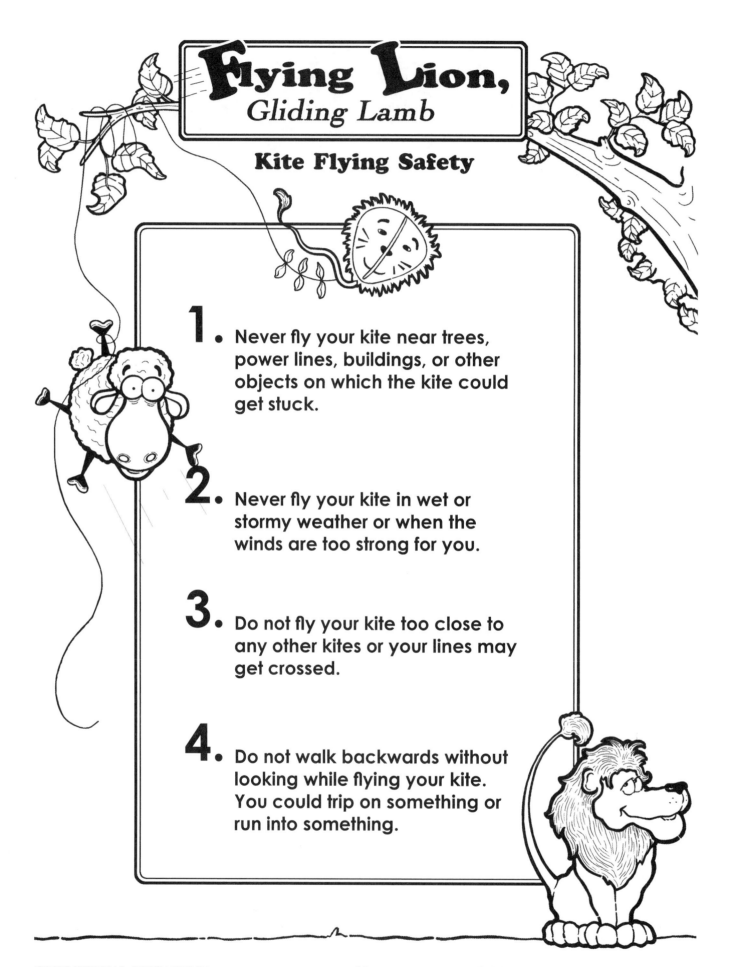

1. Never fly your kite near trees, power lines, buildings, or other objects on which the kite could get stuck.

2. Never fly your kite in wet or stormy weather or when the winds are too strong for you.

3. Do not fly your kite too close to any other kites or your lines may get crossed.

4. Do not walk backwards without looking while flying your kite. You could trip on something or run into something.

Flying Lion, Gliding Lamb
Kite Construction

1. 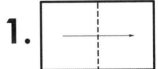 Fold your paper in half along the long edge.

2.  Line up the straight edge of the pattern with the folded edge.

3.  Trace around the pattern and cut it out.

4.  Fold the pattern and the paper along the dashed line. This folded piece will be the keel of the kite.

5. 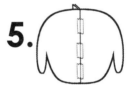 Open up the large flaps and tape along the fold you just made so that the edges line up exactly.

6. 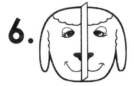 Decorate the front of the kite.

7.  Securely tape the straw to the top of the kite on the back.

8. Reinforce the area where you will be attaching the string with tape. Make a hole in the keel at the level marked on the pattern and tie the string through it. (This hole should be in the center of the keel approximately one-third of the way from the top of the kite.)

9. Attach the tail. The tail should be about three times the length of the kite body. The kite is now ready to fly!

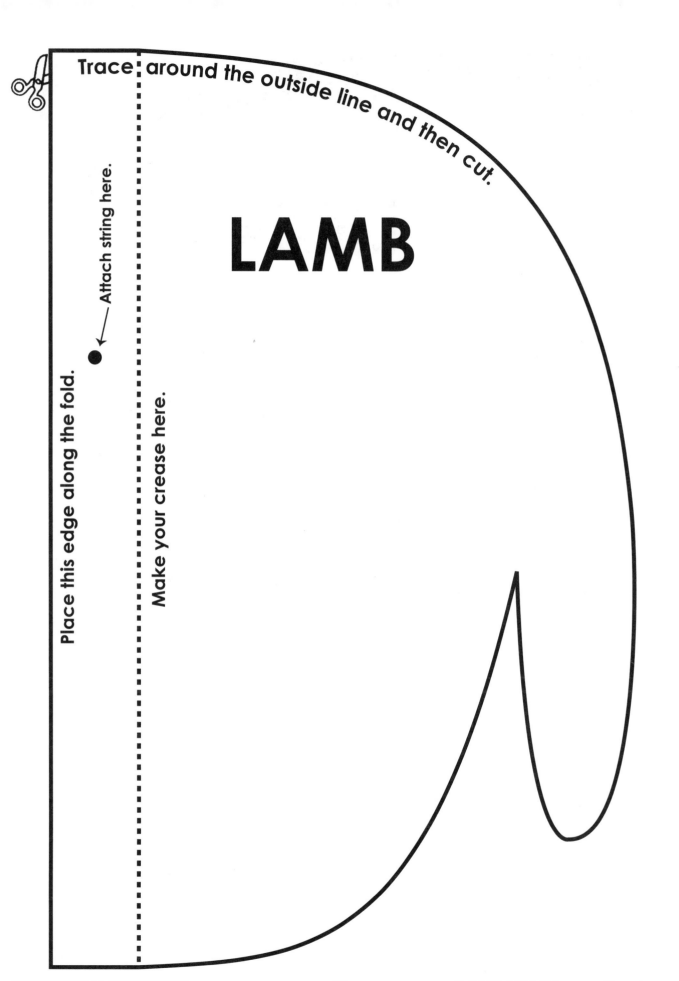

Trace around the outside line and then cut.

LAMB

Attach string here.

Place this edge along the fold.

Make your crease here.

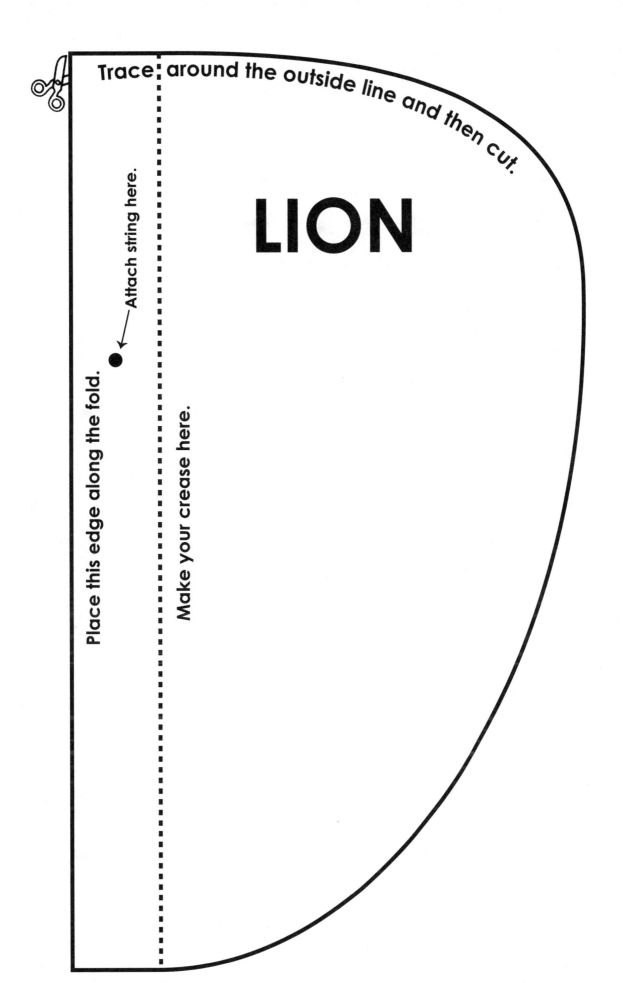

Trace around the outside line and then cut.

LION

Attach string here.

Place this edge along the fold.

Make your crease here.

Wind and Air Can Make Things Move

Tune: Twinkle, Twinkle Little Star

Do you see that kite up there? What keeps it up

in the air? High a-bove the ground it flies—

pre-tty col-ors in the skies. With-out wind the

kite can't fly. Wind's what keeps it in the sky.

See that ship go sailing by
With its sail up big and high?
On the lake it moves so fast
Just as long as wind does last.
Without wind the ship can't go.
When it's calm they have to row.

See the leaves blow all around?
They won't stay still on the ground.
See that feather in the air?
I can make it stay up there.
We have seen, we now can prove—
Wind and air can make things move.

Earth Day is on April 22nd every year. The first Earth Day was in 1970. It was started by Senator Gaylord Nelson from Wisconsin.

72

He wanted the government to think about the environment. Things needed to be done to stop pollution, clean up rivers and lakes, reduce waste, and increase recycling.

73

Now Earth Day is a yearly reminder that we need to protect our planet. There are lots of things you can do to help keep our environment healthy.

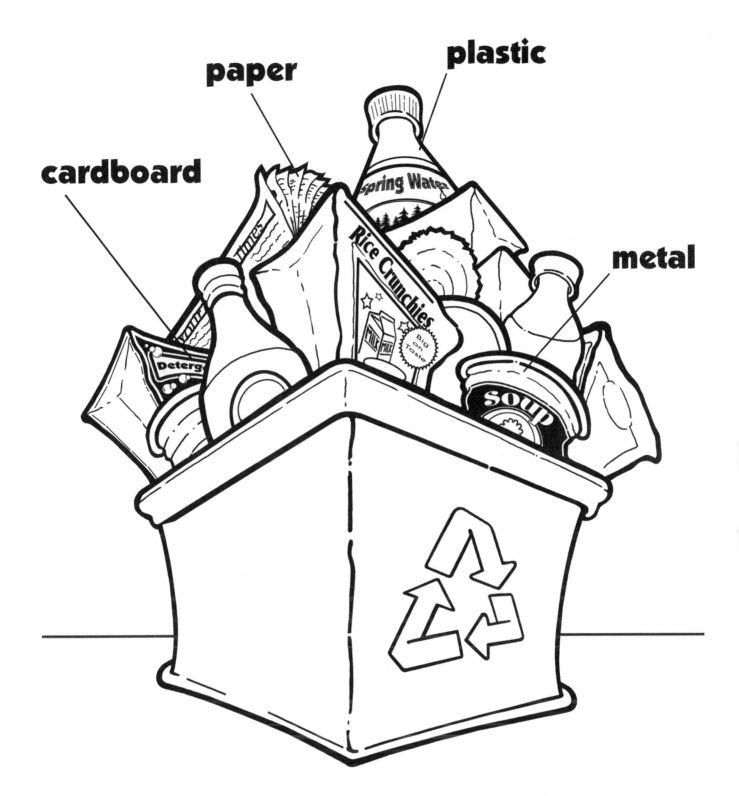

paper

plastic

cardboard

metal

One way to help is by recycling. Many things you
buy can be recycled, including paper, glass, plastic,
and metal. Go through the trash in your classroom
or at home and see if there are things that can be
recycled that you are throwing away.

Another way to help is by buying smart and reusing things instead of throwing them away. For example, buy a big bag of chips and put some in a plastic bag for your lunch. You can reuse the bag every day, and there is less trash than if you bought a small bag for each day's lunch.

You can also help the environment by reducing.
You can reduce the amount of electricity you use by
turning off the lights when you leave a room. You can
reduce the amount of pollution by riding your bike to your
friend's house instead of having your mom drive you.

Think of some ways you and our class can help the Earth. Let's start today and try to make our planet a cleaner, healthier place!

Earth Day Groceries

Participate in the Earth Day Groceries Project!

Started in 1994 by Mark Ahlness, a third grade teacher at Arbor Heights Elementary School in Seattle, Washington, Earth Day Groceries has grown to an international phenomenon.

Every year, thousands of students around the United States, and around the world get paper bags from their local grocers and decorate them with Earth Day themes. These bags are then distributed to local shoppers on Earth Day.

According to the official website, the goal of this project is: "To increase environmental awareness, to educate and empower youth to make a difference in their communities. And of course, to involve as many schools and students as possible—to start a movement, if you will!"

Getting involved is simple, and there is no cost other than the investment of your time. The Earth Day Groceries website has all the information you need to get started. It also has reports and pictures from the schools that have participated throughout the years.

Visit http://www.earthdaybags.org/ and join thousands of others around the world in raising awareness and making a positive impact in your community.

Waste Not, Want Not

Topic
Water conservation

Key Question
What are some ways you can conserve water?

Learning Goals
Students will:
- be aware of water waste in their homes, and
- learn practical ways to conserve water.

Guiding Document
NRC Standards
- *The supply of many resources is limited. If used, resources can be extended through recycling and decreased use.*
- *Resources are things that we get from the living and nonliving environment to meet the needs and wants of a population.*

Math
Whole number operations
addition
subtraction

Science
Environmental science
conservation

Integrated Processes
Observing
Collecting and recording data
Relating

Materials
2 two-gallon containers (see *Management 2*)
Water
1 set of measuring cups
Scenario cards, sets A and B
Water response cards, one per student
Waste Not, Want Not Journal for each student

Background Information
Water is wasted around the home every day. A slow-leaking faucet can waste approximately 15 gallons a day. A 15-minute shower can use up to 120 gallons of water. Letting the faucet run while you brush your teeth uses about a gallon of water. Here are some things that can be done to conserve water.
- Turn off the water while brushing your teeth.
- Wash only full loads of laundry.
- Take a five-minute shower instead of a bath.
- Wash your hands with cold water.
- Pour extra water from your glass onto a plant.
- Fix leaky pipes and faucets.
- Turn off the hose as you wash your bicycle.

This activity will let students recognize how water is wasted around the home and help them become aware of different ways they can conserve water.

Management
1. These activities are meant to be done over a number of days and can be done in small groups or as a class.
2. Label one of the containers *Fresh Water Supply* and the other *Used Water*. Fill the *Fresh Water Supply* container with 20 cups of water.
3. Prepare the sets of scenario cards in advance. They should be copied onto two different colors of card stock.
4. The *Waste Not, Want Not* water activity can be done as a whole class or at a station with adult supervision. Transparent containers are recommended so that students can see the water levels.
5. Students can complete the *Journal* activity during class time or at home. It is recommended that they have an opportunity to search around their homes for actual water conservation ideas.
6. The water response cards should be copied on colored card stock, with the smiling raindrop on one side and the worried rain drop on the other. Four sets can be made from one sheet if they are copied front to back.

80

Procedure

Part One

1. Ask students to describe instances when they've seen water being wasted. [sprinklers running too long, leaky faucets, long showers, hosing down the driveway, etc.] Tell them that they will be doing an activity that will show how water can be wasted in homes.

2. Direct students to the two containers of water. Tell them that the one labeled *Fresh Water Supply* contains 20 cups of fresh water and the other, currently empty, is for *Used Water*.

3. Tell students that the amounts of water given are not the actual amounts of water used for those tasks, but representations to demonstrate either the waste or conservation of water.

4. Have each student select either an A or B card. (Distribute all of the cards, even if some students get two.)

5. Have a student with an A card read that card aloud. Direct the student to select the correct measuring cup and scoop out the amount of water written on the card from the *Fresh Water* bucket and put it into the *Used Water* bucket. If students are unable to read the cards themselves, read the cards for them and have them follow the instructions.

6. On the chalkboard or a piece of chart paper, record the amount of water removed from the bucket each time. Continue until all of the A cards have been read.

7. Inform the class of the total amount of water that was used with the A card scenarios (see *Extensions*).

8. Repeat this procedure using the B cards. Be sure to start with one full bucket and one empty bucket.

9. Ask the students to look at the chalkboard and the total amounts of water that were used from the scenarios on each set of cards.

Part Two

1. Distribute a *Waste Not, Want Not Journal* to each student. Help the students fold their papers into fourths and have them read the pages of the booklet with you. Direct students to complete the sentences and draw in water conservation ideas. (For example, draw a person brushing his teeth with the water turned off.) Inform students that on the last page they are to choose a place in their home or yard where there is fresh water and write or draw one way to conserve water.

2. After completing the journal, allow each student the opportunity to share one conservation idea with the class. Have students share the remaining ideas with a partner.

Part Three

1. Give each student a water response card. Explain that one side represents water conservation—the smiling water drop, while the other stands for water waste—the frowning water drop.

2. Read the following scenarios. After each scenario has been read, ask students if this was an example of water conservation (smiling water drop), or water waste (frowning water drop). Have students individually make a decision and show the appropriate side of the response card.
   - There are leaves all over my street. A quick way to clean them up would be to turn on the hose and spray them into a big pile.
   - We had company for dinner last night. Our dishwasher is full, so I think I'll add dishwasher soap and turn it on.
   - I always hose down my outside windows before I wash them with window cleaner.
   - My brothers and sisters love to run through the sprinkler. They play in the water for hours!
   - Today the plumber came and fixed our leaky faucet in the kitchen.
   - We want the greenest lawn on the block so we water the grass every day!
   - I keep a pitcher of water in the refrigerator so that when we are thirsty we can have a cold drink.
   - I really want to wear my red shirt tomorrow, but it's dirty. I'm in luck! There's nothing else in the washing machine, so I'll wash this one shirt!
   - My dad takes a quick five-minute shower every morning.
   - Our sidewalk can get really dirty. I think I'll sweep off the dirt.

3. Offer students the opportunity to create scenarios for the class. Have the class respond using the same cards.

Connecting Learning

Part One

1. How was the information on the A cards different from the B cards? [A cards were about wasting water and B cards were about conserving water]
2. Which set used more water? Why?
3. How can we figure out how much more water was used by the A cards than the B cards? [subtract the B total from the A total]
4. How much more water did the A cards use?

Part Two

1. What are some ways that water can be wasted in the home?
2. How can you and your family conserve water?
3. What are some ways to conserve water at your school?
4. What are you wondering now?

Extensions

1. In *Part One*: After the students have scooped out the water according to the A cards, add the amount removed, but also ask, "If we started with 22 cups of water and we removed ___ cups of water, how much water should be left in the first container?"

2. Take the class on a tour of the school grounds. Ask them to look for water sources. Direct students to point out any water that is being wasted on campus. Have the class draft a letter to the principal or custodian noting the wasted water sources and asking if repairs or changes can be made. Ask students to include suggestions as to how water might be conserved at school.

3. Pairs of students can play a memory game by placing a set of the A and B cards face down, trying to match the water conservation card with its corresponding water wasting card. Copy the cards on the same color paper for this game.

Curriculum Correlation

Math

Give each student a sticky note. Ask them to predict the amount of water that can drip from a faucet in 15 minutes. Graph the guesses on the chalkboard. Place a pitcher under a faucet in the classroom and turn it on so that it slowly drips into the pitcher. After 15 minutes, turn it off. Using measuring cups, ask two students to measure the amount of water in the pitcher and announce the amount to the class. Have students discuss what that water could have been used for instead of being wasted. Some answers might include drink, pour on a plant, or wash hands. Use the water according to one of the suggestions.

Language Arts

Use a story starter to encourage students to write about the importance of water conservation. For example:

"It's my big brother's job to wash our dog. Yesterday I noticed that while he was scrubbing Fido out on the lawn, he left the water running. I..."

Physical Education

A relay game can be played by having students form two teams. Each team will need a tablespoon and a container of water. At the finish line there will be an empty container for each team. One at a time, a student from each team will carry a spoon filled with water to the team's cup trying not to spill water along the way. Each team should have an equal number of players. When the last member of a team empties his/her spoon of water, the game is over. The team that has the most water into their container wins. Integrate math by comparing and measuring the total amounts of water from both cups.

Literature

Asch, Frank. *The Earth and I*. Gulliver Books. San Diego, CA. 1994.

Hooper, Meridith. *The Drop in My Drink*. Penguin Books. New York. 1998.

Marzollo, Jean. *I Am Water*. Scholastic, Inc. New York. 1996.

Scenario Cards

A	I have been working in the sun. I would like some cold water to drink. I'll turn on the hose. (1 cup)	B	I have been working in the sun. I am thirsty. I'll get a bottle of water. (1/2 cup)
A	I have been playing basketball. I'll take a hot bath. (2 cups)	B	I have been playing basketball. I'll take a quick shower. (1 cup)
A	Mom asked me to wash the lunch dishes. I put them in the empty dishwasher and turned it on. (2 cups)	B	Mom asked me to wash the lunch dishes. I put them in the empty dishwasher. I'll turn it on when it's full. (1/2 cup)
A	My tennis shoes are dirty. I'll put them in the washing machine and turn it on. (2 cups)	B	My tennis shoes are dirty. I'll put them in the washing machine with my dad's work clothes. (1 cup)
A	It's hot outside. I'll play in the hose. (1 cup)	B	It's hot outside. I'll put some water in my little plastic pool. (1/2 cup)
A	We water our lawn in the afternoon when it's hot outside. (1 cup)	B	We water our lawn early in the morning. (1/3 cup)
A	I like to wash our car. I leave the water running so I can spray my sister. (2 cups)	B	I like to wash our car. I turn off the water while I wash it. (1 cup)
A	We are growing a garden. I will spray the plants every afternoon. (2 cups)	B	We are growing a garden. We use a drip hose. (1 cup)
A	I just ate a candy bar. I need to brush my teeth. I leave the water running as I brush. (1 cup)	B	I just ate a candy bar. I need to brush my teeth. I turn the water off as I brush. (1/4 cup)

Water Response Cards

Water Response Cards

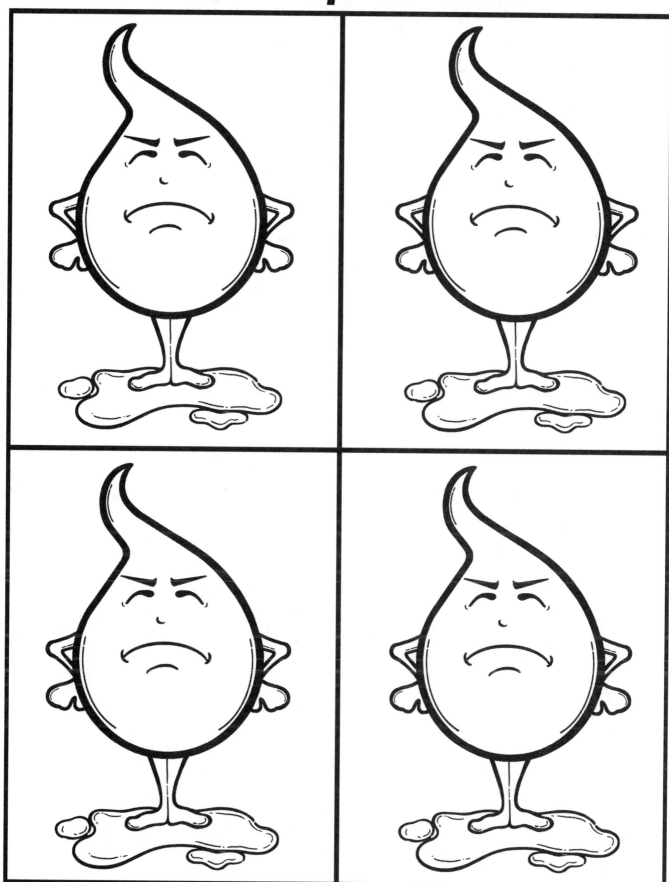

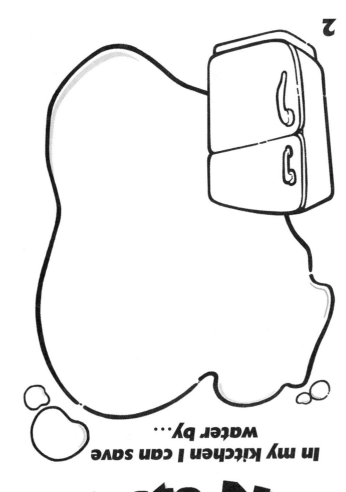

In my bathroom I can save water by…

In my kitchen I can save water by…

In my _____ I can save water by…

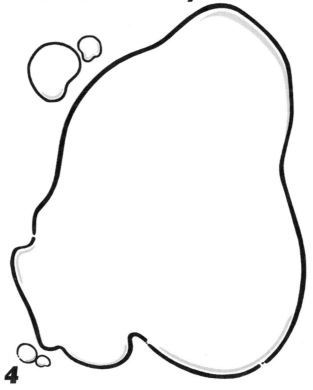

Waste Not, Want Not

Journal

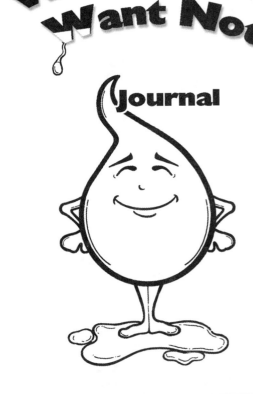

Conservationist

4

Just a Little Drip

Topic
Water conservation

Key Question
How much water is wasted from a leaky faucet?

Learning Goal
Students will construct a model with plastic cups to simulate a leaky faucet to observe how much water can be wasted.

Guiding Documents
NRC Standard
- *The supply of many resources is limited. If used, resources can be extended through recycling and decreased use.*

NCTM Standards 2000*
- *Recognize the attributes of length, volume, weight, area, and time*
- *Pose questions and gather data about themselves and their surroundings*

Math
Measurement
    time

Science
Environmental science
    water conservation

Integrated Processes
Observing
Comparing and contrasting
Communicating
Generalizing
Applying

Materials
For the class:
    one 4-foot clear plastic tube (see *Management 6*)
    funnel

For each group:
    wide-mouth, transparent 9-ounce plastic cup
    narrow-mouth, transparent 10-ounce plastic cup
    opaque 12-ounce plastic cup, 2/3 full of water
    paper towel
    sheets of newspaper

Background Information
Available water is in limited supply, yet water is constantly wasted in a home.
- Depending on the amount of water dripping, a leaky faucet may waste 15 gallons a day.
- Waiting for tap water to warm or cool can waste five to 10 gallons of water.
- Allowing the water to run while brushing teeth can waste around one gallon of water.

Younger students can help the family conserve water. They tend to be quite good at reminding the family when water is being wasted. This activity focuses on the amount of water that is wasted due to a leaky faucet. While children may not be able to fix the leaky faucet, they can bring it to the family's attention.

Management
1. Find flexible plastic cups that are easy to puncture with a pushpin. Press the pushpin into the center of the base of the wide-mouthed cup to create a small hole. It is best to puncture the hole by pushing the pin from the inside of the cup to the outside. This cup will represent the leaky faucet. The narrower-mouthed cup serves as a water catcher.
2. This activity is best done with partners so each student can observe the dripping and have an opportunity to fill the cup (turn on the faucet).
3. Place a pad of newspaper under the cups to absorb any water spillage.
4. The teacher should act as timekeeper for the first trial so the students can focus on observing the dripping. Older students will most likely be able to time their own trials. Younger students may continue to need the teacher's support in timing.
5. When collecting dripped water, the teacher should move from group to group with the large tube or graduated cylinder to collect wasted water. A funnel in the opening helps to reduce spillage.
6. The kind of long plastic tube that is recommended for this activity can be found in most large hardware stores. It is used to cover fluorescent lights. Seal one end of the tube so water cannot leak out. If this kind of tube cannot be found, graduated cylinders will also work.
7. If possible, have a faucet available for students to see. Ask the custodian if there is one that can be borrowed, or shop garage sales for an inexpensive one.

Procedure

1. Ask the class about how water is wasted around their homes. If the class does not suggest a leaky faucet, either hold up a faucet or take the class to the sink. Discuss how leaky faucets waste water.

2. Tell the students that they are going to make a device to simulate a leaky faucet using two plastic cups, one that has a hole punched in it. They will place the wide-mouthed cup (with a hole) into the opening of the narrow-mouthed cup.

3. Tell them that when you say "Begin," they will pour water from the large cup into the top transparent cup until it reaches the level of the brim of the lower cup. They will then observe how much water is wasted with just a little leaky faucet (cup with hole).

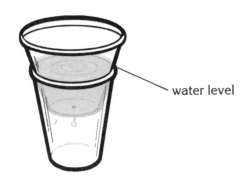

water level

4. Demonstrate how to remove the cup with the hole when time is called. Do this by lifting the cup and placing a finger over the hole. They should pour the remaining water back into the large opaque cup.

5. Explain that you will time the dripping for one minute and that they are to carefully observe the amount of water wasted in just one minute.

6. Time the first trial for one minute. Have students sketch or write about what they observed under *First I saw* on the student page. Have students hold up their collection cups so the groups can compare the amount of water that has leaked. Discuss why there may be differences.

7. Using the plastic tube (see *Management 6*), collect the water from each group. To make the students aware that this collection represents several minutes of a dripping faucet, count aloud with the students as each group pours their collected water from one minute of dripping (one minute with a leaky faucet, two minutes with a leaky faucet, etc.). As you are collecting, compare the number of minutes to an event at school with which students are familiar, such as a 15-minute recess or a 30-minute lunch.

8. Leave the accumulated water in the large tube and repeat the procedure. After each group adds the second collection of water to the tube, hold it up. Discuss how much dripping time the column of water represents. Have students record their observations in the *Then I saw* area on the student page.

9. Hold up the tube so the class can observe. Point to the water level. (Food coloring can be added to make the water easier to see.) Discuss where the water level would be for an hour ...the whole day with just a little drippy faucet.

10. Discuss what students can do if they find a leaky faucet around their house. [Tell their parents.]

11. Using the bottom of the student page, have students record ways to save water by listing or drawing them.

Connecting Learning

1. What did you learn about leaky faucets?
2. What would happen if a faucet had a bigger leak?
3. What can you do if you find a leaky faucet at home?
4. What can we do with the water wasted by a leaky faucet until it is fixed?
5. Are there other things besides faucets that leak? [hoses, pipes]
6. What are some other ways we can conserve water?
7. What are you wondering now?

Extensions

1. Puncture two or three holes and find out how much water is wasted with larger leaks.
2. Call the local water resource agency and ask for printed material or speakers on water conservation.

Curriculum Correlation

Language Arts
Cut large drops of water out of plain white paper and use it to write a story about the day the faucet dripped all day long.

Math
Allow students to use their leaky faucet device as a timer in the math classroom.

Home Link

Have students take the faucet device home and repeat the investigation with their families.

* Reprinted with permission from *Principles and Standards for School Mathematics*, 2000 by the National Council of Teachers of Mathematics. All rights reserved.

Just a Little Drip

First I saw...   Then I saw...

Ways to save water...

Topic
Reusing materials

Key Question
How can we reuse milk containers?

Learning Goal
Students will learn how to make games and puppets out of milk containers.

Guiding Documents
Project 2061 Benchmark
- *Many materials can be recycled and used again, sometimes in different forms.*

NRC Standard
- *The supply of many resources are limited. If used, resources can be extended through recycling and decreased use.*

Science
Environmental science
    reusing

Integrated Processes
Observing
Comparing and contrasting
Communicating

Materials
Half-pint milk cartons (see *Management 1*)
Plastic milk containers, one per person
Scissors
Construction paper
Glue
Decorating materials (see *Management 3*)
Balls (see *Management 5*)

Background Information
Each year Americans throw away some 400 million pounds of garbage. The United States has only 5% of the world's population and yet it produces 30% of the world's garbage. When we throw things away, they don't really "go away." Often our garbage goes into a landfill or is incinerated. Neither of these two options is without risk to our environment. The problem with incinerating trash is that the incinerator can create air pollution while eliminating our trash. Landfills can pose many problems. They take up a lot of land that could be used for other purposes and often create a toxic liquid from water percolating through the garbage in the landfill. In addition to the disposal problems, we are using up natural resources that are limited.

The most Earth-friendly solutions are to reuse, recycle, and reduce the amount of materials we use. This activity gives children two ways that they can reuse materials to make toys. Our hope is that this experience will encourage students to think of ways that they can reuse and recycle materials at home and at school.

Management
1. Provide extra milk cartons in case some are ruined in construction. Half-pint milk cartons are often available directly from a dairy if your school cafeteria uses milk bags instead of milk cartons.
2. An adult should do the cutting of the plastic milk carton.
3. Colored electrical tape, painters tape, and /or stickers work well on the plastic surface.
4. Wash the milk cartons and let them dry well before starting. Note: It may be easier to cut the bottom off the containers first.
5. For safety reasons, it is suggested that a soft ball such as a whiffle ball or rubber ball be used for the game in *Part Two*.

Procedure
Part One—Milk carton puppet
1. Give each student one milk carton.
2. Instruct the students to open the carton top.

90

3. Assist the students in cutting the two opposite corners (diagonally from each other) down to the bottom of the carton as shown in the illustration.

4. Have the students open up the cut carton by pushing in on the center bottom of the milk carton.
5. When the carton is open very wide, demonstrate how to open and close the mouth of the puppet by placing your pointer finger in the top triangle on the backside and your thumb in the bottom triangle on the backside and moving them close together and far apart.

6. When the students learn to manipulate the puppet, have them draw or glue faces of animals or people on the carton using scraps of construction paper from the recycling box. They can attach a tongue to the back of the open mouth to give the puppet more character.
7. Have the class use their puppets to tell other classes about reusing materials.

Part Two—Milk container ball catchers
1. Assist the students by cutting off the bottoms of the plastic milk cartons. Cut a U shape under the handle. Make sure that you do not cut into the handle that will be used to hold the ball catcher.

2 Encourage the students to use recycled materials, tape, and stickers to decorate their ball catchers.
3. Discuss what it means to reuse something. Ask the class what materials they reused in making this ball game. Brainstorm other games or toys that could be made from reused or recycled materials.
4. Take the class outside for a game of catch and toss with these fun toys made from reused materials.

Connecting Learning
1. What does it mean to reuse something?
2. What materials did we reuse to make the ball game? ...the puppets?
3. Why is it important to reuse and recycle?
4. What other materials could you reuse making puppets?

Extensions
1. Make sock puppets.
2. Make paper lunch bag puppets.

Curriculum Correlation
Gibbons, Gail. *Recycle*. Little, Brown and Co. Boston. 1992.

Topic
Recycling

Key Question
Out of the containers we brought to school, which plastics code will we find most often?

Learning Goals
Students will:
- observe and classify objects based on recycle codes,
- record and interpret data collected in graphs, and
- develop environmental awareness.

Guiding Documents
Project 2061 Benchmarks
- *Many materials can be recycled and used again, sometimes in different forms.*
- *Numbers can be used to count things, place them in order, or name them.*
- *Simple graphs can help to tell about observations.*

NRC Standards
- *Use data to construct a reasonable explanation.*
- *Communicate investigations and explanations.*

NCTM Standards 2000*
- *Count with understanding and recognize "how many" in sets of objects*
- *Sort and classify objects according to their attributes and organize data about the objects*
- *Represent data using concrete objects, pictures, and graphs*

Math
Counting
Graphing

Science
Environmental science
   recycling
   awareness

Integrated Processes
Observing
Predicting
Comparing and contrasting
Classifying
Collecting and recording data
Interpreting data

Materials
For the class:
   plastic containers imprinted with identification codes
   8 large paper sacks (see *Management 4*)
   prediction chart
   floor graph
   collection bin or box for recycled containers
   sentence strips for graph labels
   Just a Dream (see *Curriculum Correlation*)

For each group of students:
   grid paper for individual graphs

Background Information
Americans go through 2.5 million plastic containers each hour. At a result, landfills are rapidly filling up with tons of plastics. Because plastics do not decompose, recycling efforts are vital to the environment.

Most plastic containers are now imprinted on the bottom or side with a SPI code indicating the type of materials from which the container is made. The SPI code is established by the Society of the Plastics Industry. After collection, plastic containers are sorted by this SPI code. According to The American Plastics Council (APC) there are now more than 1400 products made with recycled plastics.

Plastic beverage containers have the second highest "scrap value" of recyclable materials. These soda bottles (PET-polyethylene terephthalate), for example, are recycled into products such as carpeting, scrub pads, fillers for jackets, and many more household and industrial uses. (Source: American Plastics Council (APC) 1-800-2-HELP-90 http://www.plasticsresource.com)

The plastics industry continues to invest in research and education to increase recycling, and students can become active participants in the recycling effort.

Management

1. One to two weeks before doing this activity, have students bring clean recyclable plastic containers from home. Because of classroom storage limitations, you may want to limit the number of objects each student brings. Provide a collection bin in which students can place their containers.

2. Copy three sets of the SPI code labels. One set will be used on the prediction chart, one will be used on the paper bags, and one will be used on the floor graph. Also copy a prediction label for each student.

3. Prepare a prediction chart for student responses using one set of SPI code labels.

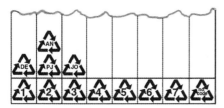

4. Tape or glue one of the SPI code labels to each of the eight paper bags.

5. Provide grid paper, one sheet per student, so that students can make their own graphs of the results of the relay.

6. Prepare an area for the relay race (inside or out).

7. This activity is divided into two parts. The first entails a relay race in which students sort plastic containers by their codes. From the relay race, they will begin to see which code appears most often. The results will be used as a basis of prediction for the larger collection of plastic containers that are classified and counted in *Part Two*.

Procedure

Part One

1. Read *Just a Dream* by Chris Van Allsburg.

2. Ask students what they know about recycling.

3. Discuss the need for recycling and how to recycle, reuse, and reduce.

4. Tell the students they are going to play a relay game to sort the recyclable plastics by their code numbers. After the game, students will make predictions as to what code number appears most often in the class collection.

5. Make certain that students recognize the SPI codes that are on the bottoms of most containers.

6. Divide the class into three groups with approximately the same number of students in each

group. If this is not possible, some students may need to run the relay twice.

7. Have each group line up behind the starting line.

8. Explain that each member will race to the collection bin, choose a plastic container, locate the code number, and place it in the appropriately numbered sack at the end of the playing field.

9. They will then race back and tag the next member in their group.

10. The game continues until all members of one team are finished, or until all of the students have run the relay.

11. Gather the bags and the collection bin and return to the classroom. Use the plastics from the bags to make a floor graph. Label and discuss the graph.

12. Distribute the grid paper to the students and have them make individual or small group graphs based on the data from the floor graph.

Part Two

1. Tell students that this is just a sample of the containers from the collection bin. Ask them to think about the code numbers on the rest of the containers and make a prediction as to which code number will appear most often.

2. Distribute the prediction labels and have students place them on the prediction chart. If desired, students can write their initials within the recycling symbol. Discuss the results.

3. Give each group approximately one-third of the containers left in the collection bin. Direct them to sort them by code numbers.

4. Bring the class together to discuss their results.

5. Have students arrange their containers on the floor graph (along with those that are already there). Ask them to compare the results with the prediction chart and relay race.

6. Have students write about their graphs.

Connecting Learning

1. Which recycling number did we find most often in the relay race? ...in the collection bin? Which number did we find the least in the relay race? ...in the collection bin?

2. How many more number____ did we find than number ____? How do you know?

3. How were your prediction graph and our total count the same? ...different?

4. Why do you think we had fewer of some numbered plastic items than others?

5. What kinds of things can we recycle?

6. What does this symbol (SPI code) mean?

7. How can we help reduce, reuse, and recycle plastics?

8. How could we use some of these plastic items?

9. What would happen if we stopped recycling?

Extensions

1. Have students look for recycling symbols on plastics they have at home. Tally the number of each type and compile the totals of all class members.
2. Design a plan for a community that does not currently have a recycling center for plastics.
3. Research what the SPI code numbers mean and why there were more of some than others.

Curriculum Correlation

Art

Using what they know about plastics, make a poster convincing others of the importance of recycling.

Make a collage or structure out of recyclable items.

Music

Make a musical instrument using recyclable materials.

Social Studies

Investigate what materials are recyclable in your town. ...in your county. ...in your state.

Take a field trip to a recycling center.

Invite a speaker from a recycling center, solid waste center, or plastics industry to the classroom.

Math

Keep a record for a week of what happens to the "throw away" containers in your own home. Tally the number of each type plastic your family discards.

Literature

The Earth Works Group. *The Recycler's Handbook.* EarthWorks Press. Berkeley, CA. 1990.

Leedy, Loreen. *The Great Trash Bash.* Holiday House. New York. 2000.

Seuss, Dr. *The Lorax.* Random House. New York. 1971.

Showers, Paul. *Where Does the Garbage Go?* Harper-Collins. New York. 1994.

Van Allsburg, Chris. *Just a Dream.* Houghton Mifflin. Boston. 1990.

*   Reprinted with permission from *Principles and Standards for School Mathematics,* 2000 by the National Council of Teachers of Mathematics. All rights reserved.

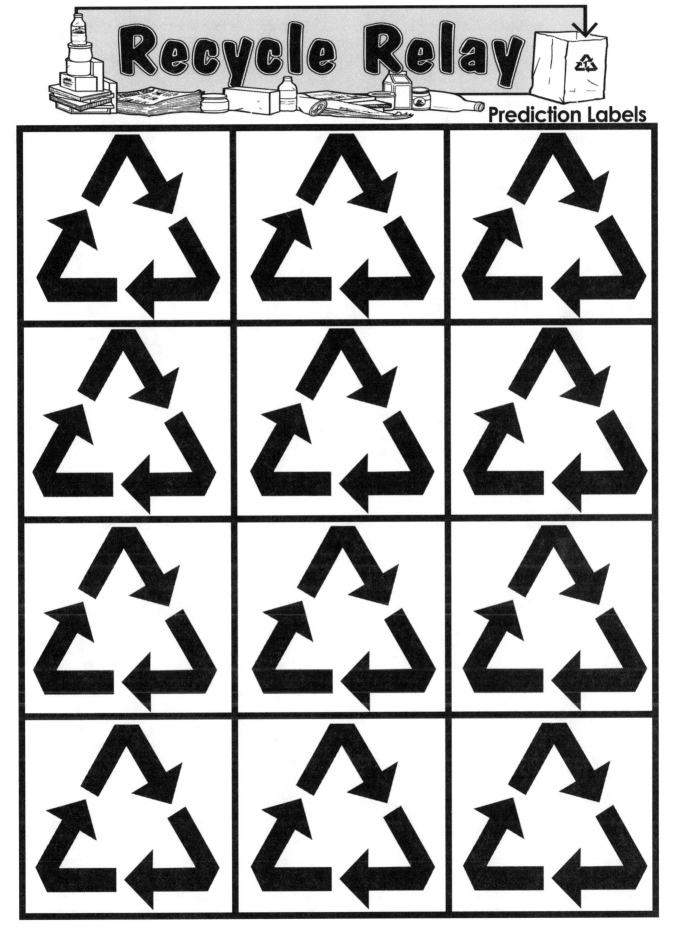

SPI Code Labels

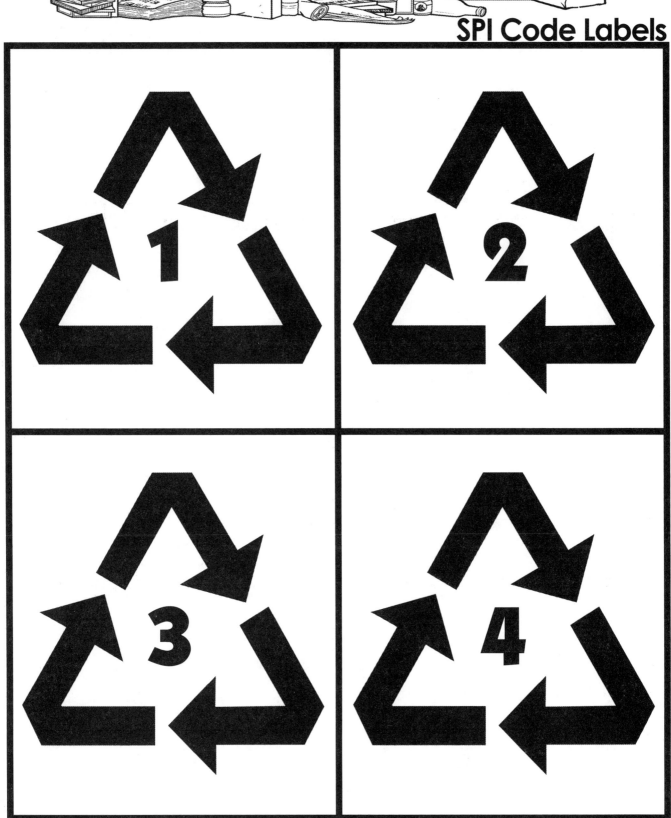

Recycle Relay

SPI Code Labels

97

Waste Watchers ♻

Topic
Reduce, reuse, recycle

Key Question
What are some ways we can reduce, reuse, and recycle?

Learning Goals
Students will:
- play a card game in which they match scenarios that show reducing, reusing, and recycling with those that don't; and
- sort the scenarios into three categories: reducing, reusing, or recycling.

Guiding Documents
Project 2061 Benchmark
- *Many materials can be recycled and used again, sometimes in different forms.*

NRC Standard
- *The supply of many resources are limited. If used, resources can be extended through recycling and decreased use.*

NCTM Standard 2000*
- *Sort, classify, and order objects by size, number, and other properties*

Math
Sorting

Science
Environmental science
   reducing
   reusing
   recycling

Integrated Processes
Observing
Comparing and contrasting
Classifying

Materials
Waste Watcher cards
The Great Trash Bash (see *Curriculum Correlation*)

Background Information
Young learners need to have an awareness of the Earth's resources and know that they are responsible for how they use those resources. Even young children can make a conscious decision to recycle a can, turn off the water while they are brushing their teeth, or reuse a paper lunch bag. This activity provides a playful context in which students can recognize Earth-friendly actions and pair those with the corresponding non-Earth-friendly actions. They will also be asked to classify the actions as reducing, reusing, or recycling.

Management
1. You will need to make one set of *Waste Watchers* cards for each group of four students. It is best if the cards are copied onto card stock and laminated.
2. This activity assumes that students are already familiar with the terms *reduce*, *reuse*, and *recycle*, and how they apply to the Earth's resources.

Procedure
1. Read *The Great Trash Bash* by Loreen Leedy. Talk about the things that the animals of Beaston did to fix their trash problem.
2. Have students get into groups of four. Distribute one set of *Waste Watchers* cards to each group.
3. Explain the rules of the game and allow sufficient time for each group to play one round.
4. After groups are finished, have students sort their pairs into three categories: scenarios that show reducing, scenarios that show reusing, and scenarios that show recycling.
5. Discuss their sorts and see if all of the groups put the same pairs in the same categories. If not, resolve the differences by having students explain their reasoning and deciding if both interpretations are justified, or if one group needs to change their sort.
6. Play the game as many times as desired.

Rules
1. Shuffle the cards and deal all of them out to the players. (In a four-person game, one player will always have one more card to start with than the other players.)

2. Each player lays down any pairs in his or her hand. Cards that show the same scenario in an Earth-friendly and a wasteful way make pairs. For example, the card showing an aluminum can in the trash goes with the card showing an aluminum can in the recycle bin.

3. The player to the left of the dealer begins. He or she can draw one card from any other player's hand. If that card is the pair to a card in his or her hand, the pair is laid down. If not, it is kept in the hand.

4. Play continues in a clockwise direction with each player selecting one card from the hand of any other player and laying down any pairs.

5. The game is over when one player is left with the *Waster* card—the only card without a pair.

6. One point is awarded for each pair of cards. The *Waster* card is worth minus five points. The player with the most points wins.

7. In the event of a tie, one bonus point is awarded for each three-pair set of cards that includes one recycling scenario, one reducing scenario, and one reusing scenario.

Connecting Learning
1. What does it mean to recycle something? What are some things we can recycle?
2. What does it mean to reuse something? What are some things we can reuse?
3. What does it mean to reduce how much you use something? What are some things we should try to reduce our use of?
4. What things do you already do to reduce, reuse, and recycle?
5. What things could you do more of or do better?

Extensions
1. Have students create their own scenarios and draw them on cards. Include these cards in the game play.
2. Remove the *Waster* card and play the game like "Go Fish."

Curriculum Correlation
Leedy, Loreen. *The Great Trash Bash*. Holiday House. New York. 1991.

* Reprinted with permission from *Principles and Standards for School Mathematics*, 2000 by the National Council of Teachers of Mathematics. All rights reserved.

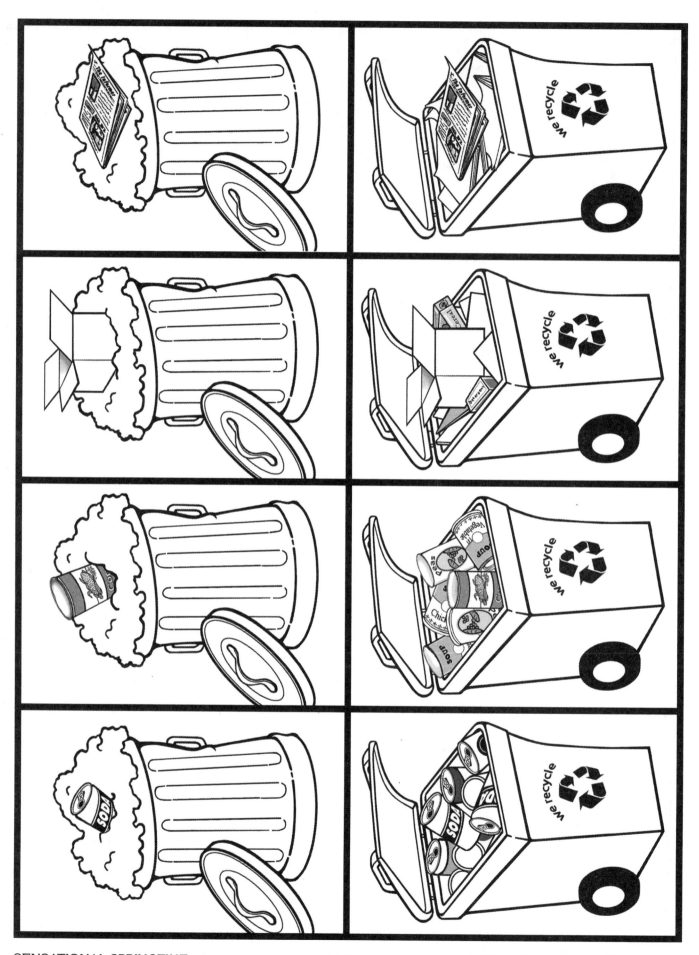

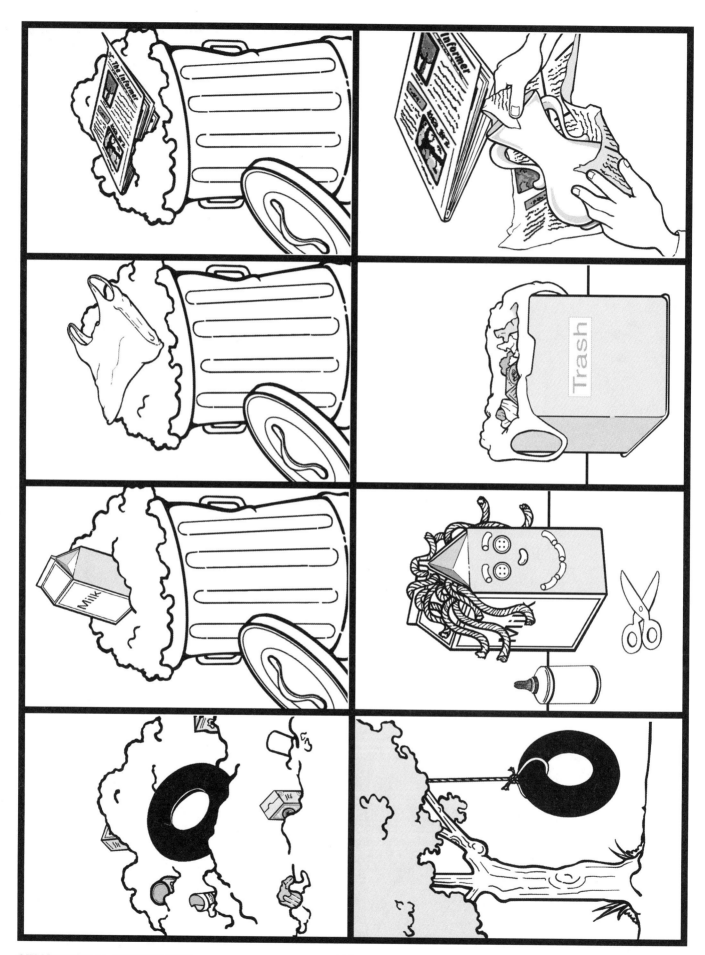

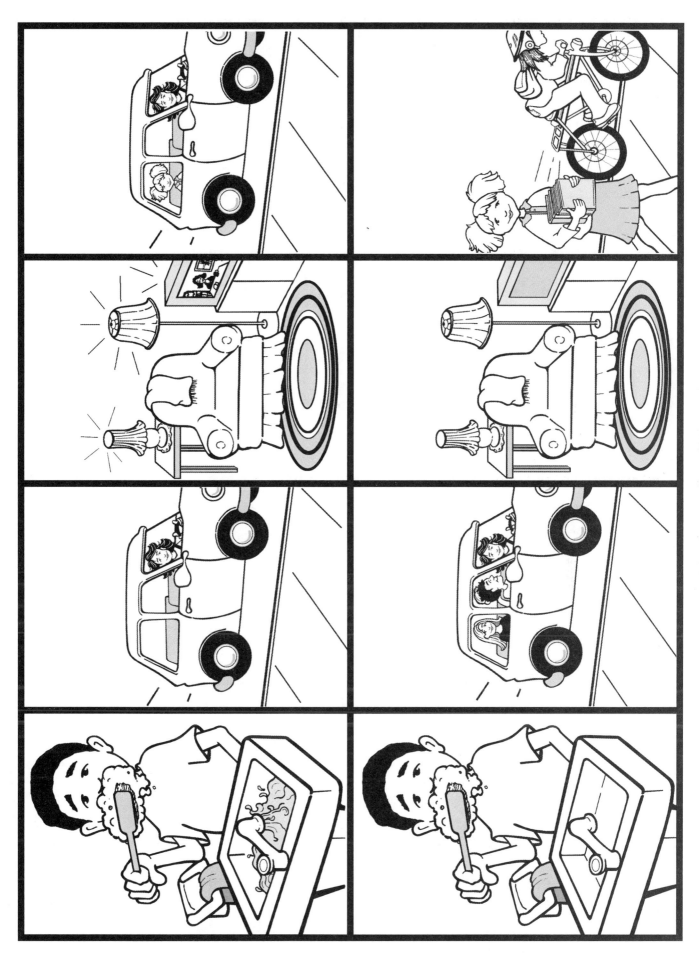

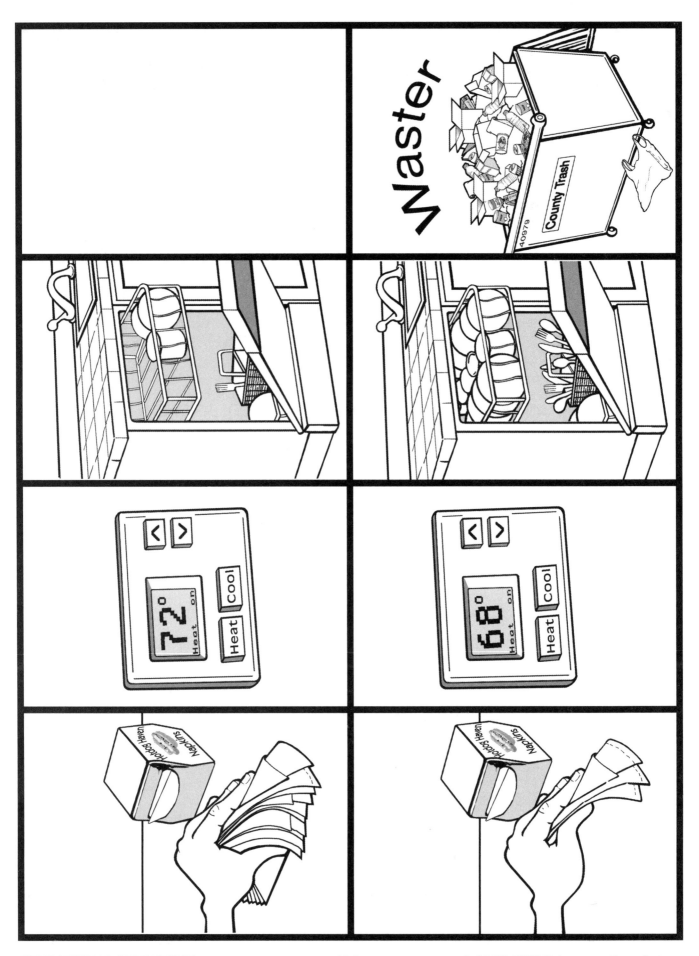

Recycle Hop

Tune: Bunny Hop
Music: Ray Anthony and Leonard Auletti

If you want to help our Earth out

there are easy things to do:

pick up cans and bottles;

Recycle, Reduce, Reuse!

Turn in cans and bottles—
You can make some money too;
It really is so simple;
Recycle, Reduce, Reuse!

Turn off the lights and water
When they are not in use;
It helps to save on power;
Recycle, Reduce, Reuse!

If you're going somewhere
That your friend is going too
You can ride together;
Recycle, Reduce, Reuse!

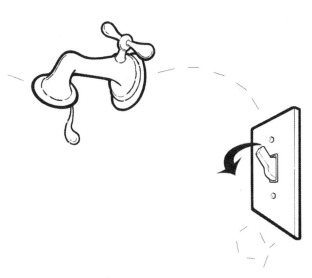

Don't throw that old shirt out
It can have another use;
It can be a dust rag;
Recycle, Reduce, Reuse!

If you have some old toys
That you no longer use,
Give them to a neighbor;
Recycle, Reduce, Reuse!

These are only some ways
To help our Earth renew
Can you think of others?
Recycle, Reduce, Reuse!

Topic
Graphing

Key Question
How can making a graph help you compare objects?

Learning Goal
Students will make a graph to show the number of insects in each garden.

Guiding Document
NCTM Standards 2000*
- *Represent data using concrete objects, pictures, and graphs*
- *Describe parts of the data and the set of data as a whole to determine what the data show*
- *Count with understanding and recognize "how many" in sets of objects*

Math
Counting
Graphing
Equalities and inequalities

Integrated Processes
Observing
Comparing and contrasting
Classifying
Collecting and recording data

Materials
Crayons
Student pages
Pasta (see *Management 2*)
Rubbing alcohol (see *Management 3*)
Food coloring (see *Management 3*)
Transparency (see *Management 4*)

Background Information
Graphs are organizational tools that are very appropriate for young children to use. However, for organizing data to be relevant, there needs to be a purpose for organizing. Comparing and counting activities can provide purpose for organizing and can help to develop children's understanding of graphing and analysis of data.

In this activity, the children will be using pasta as a manipulative on a garden work mat. They will then use the pasta to create concrete graphs to assist them in answering a series of questions. Eventually, they will remove the pasta from the grid and color in one box for each insect to make an abstract record of their insects for analysis.

Management
1. Copy one set of student pages for each student.
2. Purchase two boxes of bowtie pasta (farfalle) and two boxes of corkscrew pasta (cavatappi). If pasta is not available, you can copy the page of butterflies and caterpillars provided.
3. To turn the pasta into colorful caterpillars and butterflies, place a small amount of rubbing alcohol and a few drops of food coloring into a container, add the pasta and stir until they reach the color intensity desired, then place them on newspaper to dry.
4. Copy the first student page onto an overhead transparency.
5. Work through several problems before asking students to color the graph.
6. In the process of placing the insects in the garden, the student are also reviewing directional or positional words such as *above, to the left of,* etc.
7. To check for understanding and to move the students to an abstract level of graphing, an additional student page is included that has pictures of butterflies and caterpillars for the students to count and graph. This page is optional.

Procedure
1. Place the transparency of the first student page on the overhead. Explain that you will be using pasta to represent butterflies and caterpillars in the garden.
2. Invite a student to place one caterpillar beside the tree on the left and one beside the tree on the right. Direct another student to place three butterflies in the garden above the flowers on the right. Ask the class if you have more butterflies or caterpillars in the garden.
3. Remove the butterflies and place one in the first box to the right of the butterfly picture. Place the additional butterflies in the next two boxes.

4. Remove the caterpillars and place them in the boxes to the right of the caterpillar picture.

5. Tell the class that you have just created a graph to show you what insects were in the garden. Again ask the class which insects you have more of. [butterflies] Ask how many more butterflies you have than caterpillars. Demonstrate how to count on to determine how many more you have.

6. Give each child five pieces of bowtie pasta, five pieces of corkscrew pasta, and a copy of the first student page.

7. Direct students to place two butterflies in the treetops and one caterpillar below the trees on their student page. Ask which insect they have more of in their gardens and how many more they have of that insect. If students do not naturally move the objects down to the graph, suggest that they make a graph to show what is in their garden.

8. Discuss the graph. Ask how many insects they have in their garden all together, which they have more of, how many more they have, etc.

9. Continue asking the students to place specific numbers of butterflies and caterpillars in the garden and to compare the amounts by creating concrete graphs.

10. After students have had several opportunities to build concrete graphs, give them an additional problem to display in the garden and graph. When the students have the correct number of butterflies and caterpillars on the graph, ask them to remove the pieces of pasta one at a time and color the boxes beneath them to make a bar graph. Discuss the results.

11. If desired, distribute the second student page and have students make a graph based on the picture.

Connecting Learning

1. How many caterpillars are in the garden?
2. How many butterflies are in the garden?
3. Which insect do you have more of? How many more do you have?
4. Which insect do you have less of? How many less do you have?
5. How many insects were in the garden all together?
6. How did making a graph help you answer these questions?

Extension

Add orzo to represent eggs and shell pasta to represent the chrysalis in the garden. This will allow you to address the different stages of the life cycle of a butterfly.

* Reprinted with permission from *Principles and Standards for School Mathematics*, 2000 by the National Council of Teachers of Mathematics. All rights reserved.

Garden Graphing

Which do you have more of? _____

_____

Garden Graphing

Which do you have more of? _____

Garden Graphing

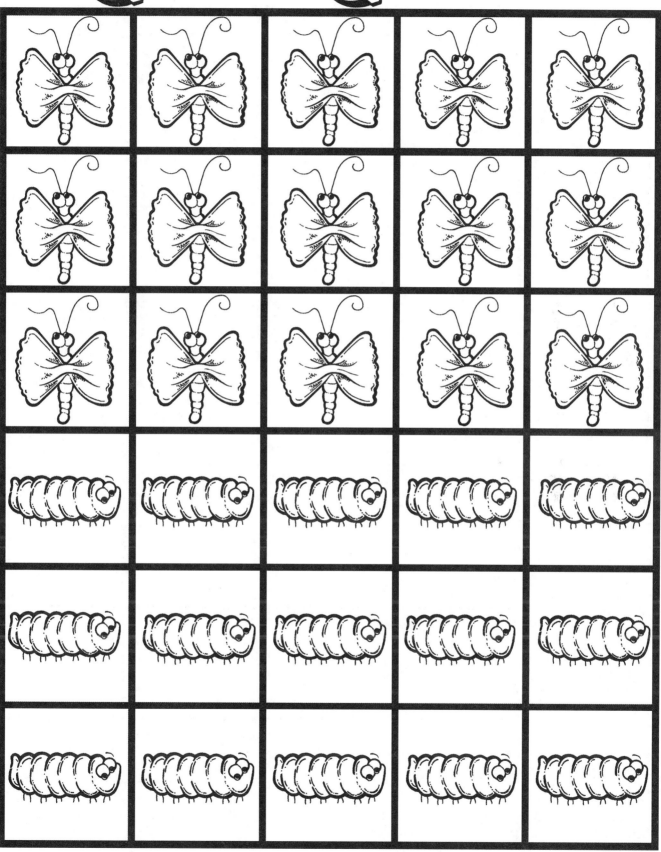

Stamping Into Spring

Topic
Sorting

Key Questions
1. How many different ways can postage stamps be sorted into groups using various features or characteristics? (color, faces, buildings, words, animals, etc.)
2. How can we use Venn Diagrams to organize postage stamps?

Learning Goal
Students will sort postage stamps according to different attributes using a Venn diagram.

Guiding Document
NCTM Standards 2000*
- *Sort and classify objects according to their attributes and organize data about the objects*
- *Represent data using concrete objects, pictures, and graphs*

Math
Data analysis
    sorting
    Venn diagram

Integrated Processes
Observing
Communicating
Classifying
Comparing and contrasting
Collecting and recording data

Materials
A variety of postage stamps (see *Management 2*)
Plastic bags, pint size
Hand lenses
Chart paper (see *Management 3*)
Grouping circles (see *Management 4*)
Labels for circles
Recording page

Background Information
Classification is an important skill for students to practice in a variety of contexts. Through sorting and classifying activities, children learn and use critical thinking skills. The ability to see and describe attributes will carry throughout the curriculum. For math and science, these skills allow students to group and regroup manipulatives and ideas. In language arts, the use of attributes allows students to better describe objects and situations.

Young children need a variety of opportunities to sort and classify in structured and in unstructured situations. As the students explore materials, they will see different relationships and develop needed vocabulary to describe these relationships. Stamps provide an interesting vehicle for students to use for these important sorting and classifying skills. They can be used over and over as students gain experience and skill using process skills.

Management
1. This activity is divided into three parts. *Part One* has students compare and contrast stamps. *Parts Two* and *Three* utilize Venn diagrams for classification.
2. You will need a large collection of stamps for this activity. You may use the stamps provided, but you should also supplement this selection with real stamps. See *Resources* for places to purchase inexpensive postage stamps.
3. Prepare several pieces of chart paper—some with one-circle and others with two-circle Venn diagrams—for recording students' sorting results.
4. Grouping circles are available from AIMS. If you do not have grouping circles, yarn circles or circles drawn on chart paper will work. It is suggested that you use two colors for the Venn circles so that students can see the distinction between sets.
5. Copy the page of labels to record the attributes the students generate for use in the Venn diagrams.
6. Each group of three or four will need a plastic bag filled with 10 to 15 stamps for *Part Three*. These stamps can be the same ones used in *Parts One* and *Two*, or they can be different.

Procedure
Part One
1. Spread out a collection of stamps and allow the students to use hand lenses as they observe and explore the illustrations on the stamps.
2. Working with partners, have students compare and contrast the collection before doing the more formal activity of coming up with a classification system.
3. Have all groups share their observations with the rest of the class.

Part Two

1. Put one grouping circle on the floor or a table. Spread out the stamps and have students observe them.
2. Working as a whole class, have the students come up with a rule for sorting the stamps; for example, stamps with animals.
3. After a rule has been identified, invite the students to work together to put all of the stamps with that attribute into the circle.
4. Write the attribute on one of the circle labels and label the group of stamps.
5. Identify the group of stamps that is not in the circle as being in the no/not group (for example, faces on stamp, no faces on stamp, etc.).
6. Continue to sort the stamps in this manner until the students are comfortable with using the rules for this type of sorting.
7. After students are comfortable with the single circle, add a second separate circle and use two rules that have already been used, for example, things that fly and animals. Have students sort the stamps into the two circles, again labeling each circle.
8. Students should notice that some stamps need to be in both circles. When students begin to see this, overlap the two circles, creating a third region where they intersect.
9. Explain to the students that sometimes a stamp has both attributes and needs to be in a space that will show that it belongs in both circles.
10. Have students continue to sort the stamps into a two-circle Venn diagram using various rules they decided on earlier.
11. Record several of the sorts developed by the students using the charts suggested in *Management 3*.

Part Three

1. When the students are comfortable with the process of sorting into two-circle Venn diagrams, divide them into small groups of three or four students.
2. Give each group a plastic bag filled with stamps and several copies of the Venn diagram page. Allow time for groups to sort their bags of stamps and make records of their sorts.
3. After a discussion about the sorting and classifying of the stamps, let students design their own stamp. Have them think about who or what they want to honor and the value of the stamp.

Connecting Learning

1. In what ways are the stamps alike? In what ways are they different?
2. What were some of the ways that the class decided to sort the stamps?
3. How many different ways did your group find to sort your set of stamps?
4. How were your sorts the same as those of other groups? How were they different?
5. How did you decide which stamps belonged in the intersection of the two circles?
6. How many different ways do you think there are to sort the stamps?
7. If I opened a new bag of stamps, what are some of the ways you think you would be able to sort them? Why?
8. If you could change one thing about our stamp sorting, what would you change? Why?
9. What are you wondering now?

Extensions

1. Determine the number of stamps it would take to do a variety of tasks—make a border around the outside of a desk, cover the front of a textbook, etc.
2. Encourage students to ask questions that can be answered using their stamps. Have them set up experiments to help them find the answers to their questions.
3. Have students make up math problems using the values listed on the fronts of the stamps and trade them with other students to complete.
4. After small groups of students have sorted the stamps using Venn diagrams, have other groups come and see if they can tell what the rules for sorting were.
5. If foreign stamps are used, find out their origin and put a pin in the world map to show that area.
6. Visit a post office and see how they use stamps.
7. Write a letter to the post office asking questions of interest to the students.

Resources

1. Kenmore Stamp Company
   http://www.kenmorestamp.com/
   (Purchase 500 worldwide stamps for just $2.00. Click on the "Great Bargains" link.)
2. Hobby Supplies
   http://www.hobbysupplies.com/
   (Packets of US and worldwide stamps available starting at $1.99. Click on the "US Stamps" or "WW Stamps" links.)
3. Universal Philatelic Service
   http://stores.ebay.com/UNIVERSAL-PHILATELIC-SERVICE
   (Wide variety of stamps available from many sources in many different categories.)

* Reprinted with permission from *Principles and Standards for School Mathematics*, 2000 by the National Council of Teachers of Mathematics. All rights reserved.

Labels for My Circles

Stamping Into Spring

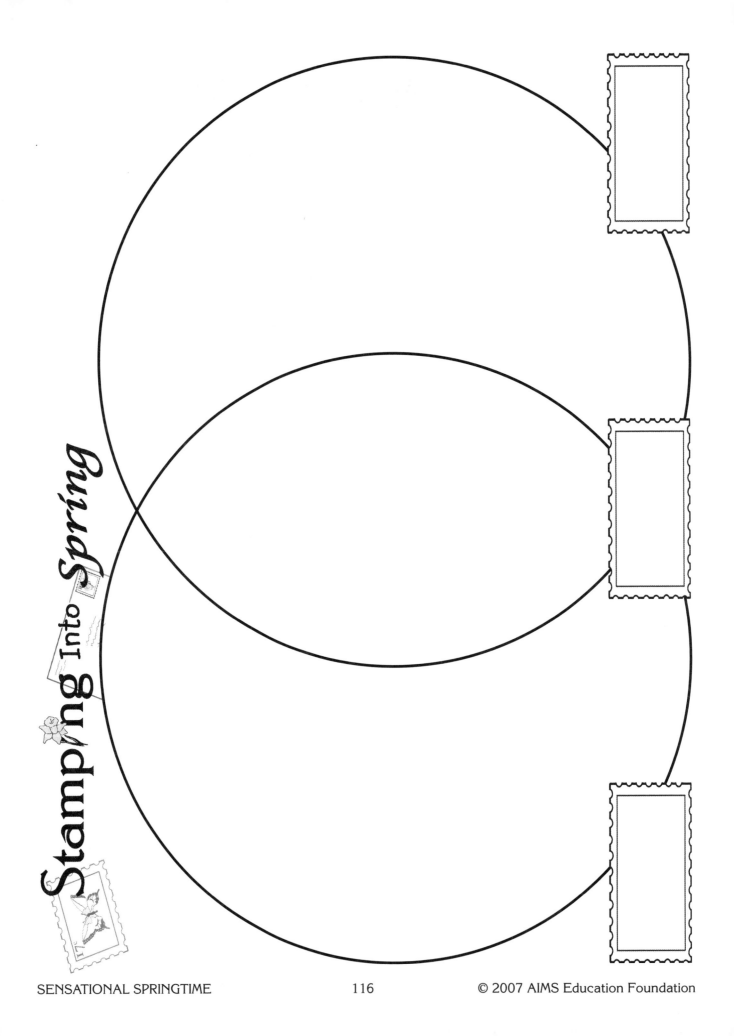

Stamping Into Spring

Cereal Sorters

Topics
Data analysis
Patterns

Key Questions
1. How can you sort your cereal sample by shape?
2. What different patterns can be created using cereal?

Learning Goals
Students will:
- sort and classify a sample of cereal with different shapes;
- graph the results of their sorts; and
- recognize, create, and extend patterns using cereal.

Guiding Documents
Project 2061 Benchmarks
- *Numbers and shapes can be used to tell about things.*
- *Numbers can be used to count things, place them in order, or name them.*
- *Simple graphs can help to tell about observations.*

NCTM Standards 2000*
- *Sort, classify, and order objects by size, number, and other properties*
- *Sort and classify objects according to their attributes and organize data about the objects*
- *Represent data using concrete objects, pictures, and graphs*
- *Recognize, describe, and extend patterns and translate from one representation to another*

Math
Sorting
Graphing
Patterns

Integrated Processes
Observing
Comparing and contrasting
Classifying
Recording
Communicating

Materials
One box of cereal that includes different shapes
Portion cups
Sorting mats (see *Management 1*)
Cereal graphs (see *Management 2*)
Glue
Pattern strip holder, one per student
Pattern strip, one per student
Scissors
Crayons
Card stock
Tape

Background Information
Cereal can make an interesting and inexpensive tool for children to use to sort and classify, graph, and create patterns. This activity can be accomplished using any cereal that includes varied shapes. This activity is divided into two parts and should be extended over a period of several days. The first part includes the sorting and graphing using a small portion of cereal. In *Part Two*, students use some of the same cereal to create and extend patterns and glue the cereal onto pattern strips.

Management
1. Duplicate one sorting mat per student or group of students.
2. Before making copies, label the columns of the graph with appropriate labels matching the shapes of cereal student will be sorting. Duplicate one graph on card stock per student or group of students.
3. Copy the pattern strip and pattern holder pages onto card stock.
4. To make the pattern holder, cut along the solid lines and fold along the dashed lines. Fold the short tab in first, and then tuck the large tab behind it. Tape down the short tab. This creates a space in the pattern holder to allow room for the cereal.
5. Divide the contents of the box of cereal equally among the groups.

Procedure

Part One—Sorting

1. Show the class a sample cup of cereal to be used for the activity and discuss the shapes that are included.
2. Ask the students how many of each shape they think are in the cup and for ideas of how they could find out.
3. Have students get into groups, if desired, and distribute the sorting mat. Give each student/group a portion cup of cereal.
4. Discuss how the mat will be used to help organize the cereal and answer the question of how many of each shape are in the cup.
5. Direct the students to sort their portions of cereal using the sorting mat. Discuss the results of the sort.
6. Distribute the graph page to the groups/students and model how they will build a real graph using the cereal.
7. Once the students have placed the cereal shapes into the correct columns on the graph, instruct them to glue the cereal onto the graph.
8. Ask the students to discuss the results of their individual (or group) data compared to other students' data.

Part Two—Patterns

1. Show the class a small portion of the cereal to be used in the activity.
2. Ask the students to discuss some possible ways they could use the cereal to create various patterns.
3. Model gluing a cereal pattern onto a pattern strip to make a permanent record of the pattern.
4. Distribute cereal, a pattern strip, and a pattern strip holder to each student.
5. Direct the students to create and glue down their own cereal patterns.
6. Once the students have completed their individual pattern strips, model making and using the pattern strip holder to reveal a pattern one shape at a time. To make it easier to slide the pattern strips, gently squeeze the sides of the pattern holder when you are inserting a strip or pulling it out.
7. Have students practice predicting which shape will come next in the pattern.
8. Ask students to exchange their pattern strips and holders and explore and predict the various patterns.

Connecting Learning

Part One—Sorting

1. Which shape did you have the most of in your cereal sample?
2. Did you have some of each shape? Did you have the same number of each shape?
3. Did each person/group have the same type of cereal most often? Why or why not?
4. How was your graph the same as your classmates'? How was it different?
5. If you were given another sample of the cereal, is there any way to predict what shape you would have the most of?
6. Why would it be important to know what shape of cereal is most common in a box of cereal?

Part Two—Patterns

1. How did you decide what pattern to make with your cereal?
2. Was your partner able to repeat and continue the pattern that you created?
3. Were there any patterns that had more than one way to extend them? Give an example.
4. What would the pattern you created look like in the everyday world? How about your partner's pattern?
5. How do you use patterns at home? How do you use patterns at school?

* Reprinted with permission from *Principles and Standards for School Mathematics*, 2000 by the National Council of Teachers of Mathematics. All rights reserved.

Cereal Sorters
Sorting Mat

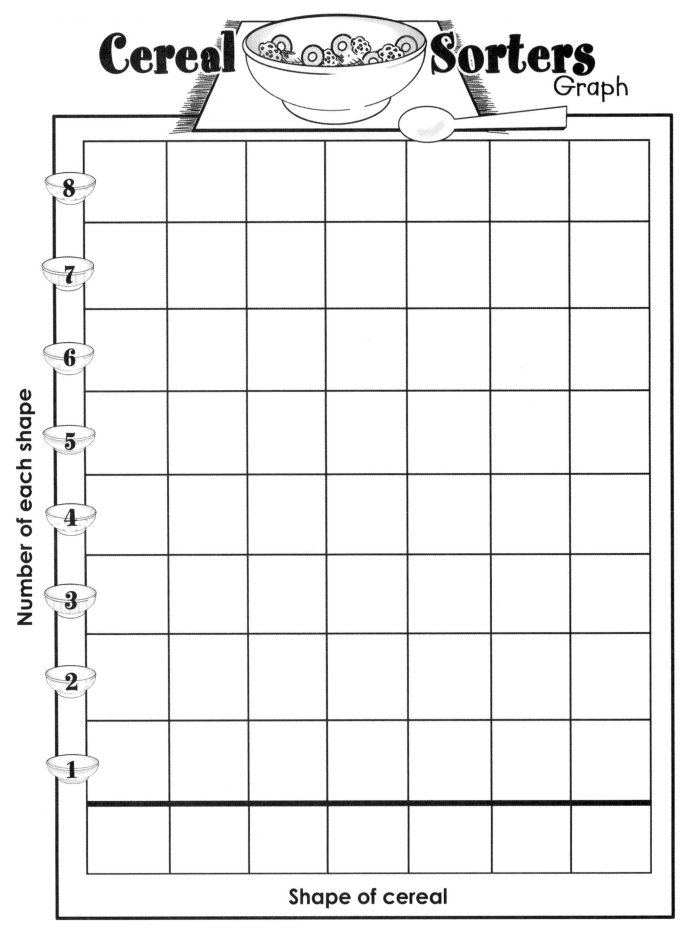

Cereal Sorters Graph

Number of each shape

8
7
6
5
4
3
2
1

Shape of cereal

Cereal Sorters
Pattern Strips

Copy onto card stock and cut out one pattern strip for each student.

My Cereal Pattern

My Cereal Pattern

Cut along the
solid lines.

Fold along the
dashed lines.

(Tuck this edge under the short flap and tape together.)

My
Pattern
Holder

Name:

Eggsploration Stations

Topic
Process skills

Key Question
What can you find out about an egg by observing and measuring it?

Learning Goals
Students will:
- observe and explore some properties and attributes of eggs; and
- use non-standard units to measure eggs in a variety of ways including mass, circumference, and volume.

Guiding Documents
Project 2061 Benchmarks
- *People can often learn about things around them by just observing those things carefully, but sometimes they can learn more by doing something to the things and noting what happens.*
- *Tools such as thermometers, magnifiers, rulers, or balances often give more information about things than can be obtained by just observing things without their help.*

NRC Standards
- *Ask a question about objects, organisms, and events in the environment.*
- *Employ simple equipment and tools to gather data and extend the senses.*
- *Simple instruments, such as magnifiers, thermometers, and rulers, provide more information than scientists obtain using only their senses.*

NCTM Standards 2000*
- *Recognize the attributes of length, volume, weight, area, and time*
- *Understand how to measure using nonstandard and standard units*
- *Use tools to measure*
- *Pose questions and gather data about themselves and their surroundings*
- *Represent data using concrete objects, pictures, and graphs*

Math
Measurement
    mass
    length
Data collection

Science
Life science
    eggs

Integrated Processes
Observing
Predicting
Collecting and recording data
Comparing and contrasting
Communicating

Materials
For each student:
    recording journals
    one hard-boiled egg
    one raw egg

Station One: Spinning Eggs
    plastic plate labeled "Hard-boiled"
    plastic plate labeled "Uncooked"

Station Two: Finding the Mass of an Egg
    balance
    non-standard units of mass (counters, Unifix cubes, etc.)

Station Three: Finding the Volume of an Egg
    9-ounce clear plastic cup
    container of water
    newspaper
    paper towels
    permanent marking pen

Station Four: Finding the Circumference of an Egg
    string
    prediction graph (see *Management 6*)
    scissors
    glue

Station Five: Rolling Eggs
    chart paper rolling area (see *Management 7*)
    paper egg markers, one per student
    graph (see *Management 8*)

Station Six: Dyeing an Egg
    old nylon hosiery (8-10 pairs)
    egg dye (see *Management 9*)
    spoon
    fern or parsley sprigs
    twist ties
    paper towels
    glue
    disposable plastic cups

Station Seven: Observing and Sketching an Uncooked Egg
    hand lenses
    9-ounce clear plastic cup

Station Eight: Cooking "Bunny in a Hole"
    electric griddle or fry pan
    biscuit cutter or round cookie cutter
    spatula
    butter
    plastic knives
    forks
    plates
    bread, one slice per student

Background Information

This activity provides the opportunity for students to practice a variety of valuable skills in math and science by exploring eggs. Students will find the masses of their eggs and cut strings equal to the circumferences. They will record an indirect measure of volume by seeing how much water an egg displaces. They will make and test predictions and practice their observation skills as they look at the parts of a raw egg and compare those to the same egg when cooked. Students will also learn to keep careful records of their observations in their journals.

Management

1. Send home a letter several days prior to this activity asking parents to send at least one hard-boiled egg and one raw egg to school with their children. Ask parents to mark the raw egg by writing an "R" on the shell with a permanent marker. Have a few extras of each kind on hand in case a child forgets or an egg is broken during station time.

2. This activity can be done in one long period or over several days with stations remaining set up the entire time. If you do the activity over several days, be sure to store the eggs in the refrigerator each night.

3. It is helpful to recruit parent volunteers or cross-age tutors for the days you do these stations. It is essential that you have adult supervision at the cooking station and the egg-dyeing station. An adult or older student can also help students to make more careful observations and sketches to help them when learning about the parts of a raw egg at station seven.

4. The students can rotate through the stations in partners or small groups.

5. Copy the station cards and put them at each station along with the materials. To prepare the recording journals, copy the student pages, cut them in half, put them in order, and staple along the left edge. Each student needs his or her own journal.

6. The circumference prediction graph for station four should be made ahead of time by cutting three strips of masking tape each about 50 cm long. Attach these strips to the wall, sticky side out, by placing a small strip of tape (approximately 4 cm) on each end of the long strips. Place the labels *Too long, Too short,* and *Just right* above each strip of tape. The students can then stick their predicted circumference strings on the sticky tape with the appropriate label.

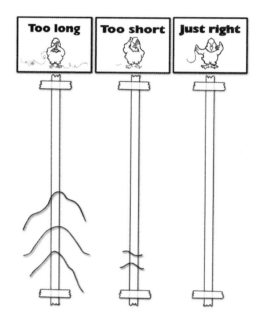

7. To make the rolling surface for station five, divide a piece of chart paper into four equal sections. Color each section a different color (or write the name of a different color in each section). Tape the surface to the table and mark a starting spot from which students will roll their eggs.

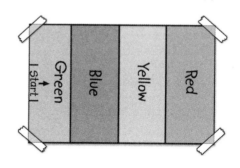

8. To make the graph for station five, prepare a piece of chart paper labeled with the four colors that correspond to the sections on the rolling surface. Title the graph "My egg stopped on."

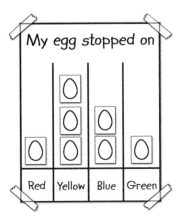

9. For station six, you can purchase commercial egg dye or make your own. To make your own, obtain skins from six to eight red Bermuda onions. (If you ask the produce manager at your local grocery store, he or she will often save them for you.) Place the skins in a pot with one quart of water. Boil the water and skins for approximately 10 minutes. Strain and save the water. Add one tablespoon of vinegar to the natural onion skin dye. This dye works effectively when it is used at room temperature.

10. Stations one to five can be done in any sequence with students moving to a station that has available space. Stations seven and eight must be done in order because the egg that students crack and sketch in station seven is the egg used to cook in station eight. Station six can be done before or after stations seven and eight, but should be done after the first five stations because it involves dyeing the hard-boiled egg that they will use in stations one to five.

11. It is assumed that students know how to use all of the tools that will be used in each center. If students are unfamiliar with any, introduce the tools separately and then use them in the center.

Procedure

1. Prior to starting the stations, gather students together and hold up an egg. Ask them to tell you what they know about eggs. Record their ideas on a large egg-shaped chart. Tell them that they are going to learn more about their own eggs and that they will be adding information to the chart later.

2. Have students gather around the first station. Explain and model the procedure to be done at that station.

3. Continue through the stations explaining and modeling the procedure at each one.

4. Assign students to their pairs or groups and give each student a journal and their eggs. Be sure that the raw eggs are marked to distinguish them from the hard-boiled eggs. (If any students get their eggs mixed up before they are labeled, simply spin them to determine which is raw and which is hard-boiled. The hard-boiled egg will spin much quicker and longer than the raw egg.)

5. Since they only need the raw eggs at stations one, seven, and eight, have them leave the raw eggs at their desks except when they are being used.

6. Have students proceed to a station, work at their own pace, then move on to another station. Continue this procedure until everyone has completed all tasks, or until the time allotted for the day is up.

7. At the end of the work time, have students put their station journals away and put their eggs on a tray to be taken to the refrigerator.

8. After all students have had a chance to visit all of the stations, review what they have learned about eggs and add the information to the chart that you started at the beginning of the activity.

9. Review individual student's journal responses. Compare observations and discuss the results of each station investigation.

Connecting Learning

1. What did you observe about the eggs when you tried to spin them? Which spun better? [the hard-boiled egg] Why do you think this happened?

2. Did all of the eggs have the same mass? Why do you think there may have been differences?

3. What happened to the water in the plastic cup when you put the egg into it?

4. Was your guess for the circumference of your egg too long, too short, or just right?

5. How far along the racing surface did you roll your egg? Did everyone in your group roll to the same place? Why do you think there was a difference?

6. Do you think you would have had the same results if you had rolled your raw egg instead of your hard-boiled egg? Why or why not?

7. Did all of the eggs turn out the same after you dyed them? What differences did you notice? Why do you think this happened?

8. What were the parts of the uncooked egg that you noticed? Did all of the uncooked eggs look the same?

9. How did the uncooked egg change when you made the "bunny in a hole"? Did any of the uncooked parts of the egg remain the same?

10. What are you wondering now?

* Reprinted with permission from *Principles and Standards for School Mathematics,* 2000 by the National Council of Teachers of Mathematics. All rights reserved.

Eggsploration Stations

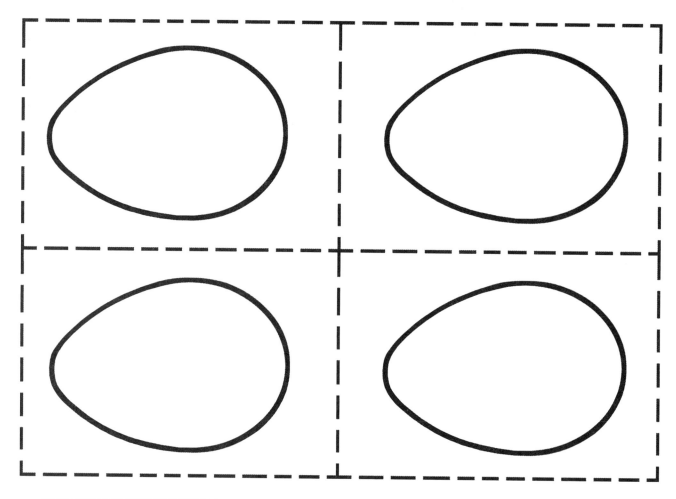

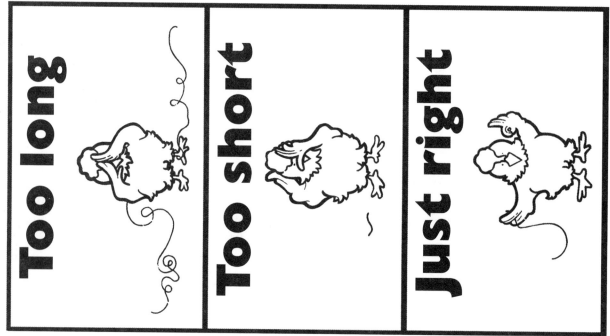

Which egg will spin easier, the hard-boiled or uncooked?

1. Place your hard-boiled egg in the plastic plate labeled *hard-boiled*.
2. Give it a spin and observe.
3. Spin the uncooked egg on the other plate and observe it.

What is the mass of your egg?

1. Record your prediction.
2. Put your egg on one side of the balance.
3. Add masses to the empty basket until it balances with the egg.
4. Record the actual mass of your egg.
5. How does the actual result compare with your prediction ?

SENSATIONAL SPRINGTIME   127   © 2007 AIMS Education Foundation

What do you think will happen to the level of water in the cup when you place your egg in it?

1. Observe the level of water in the cup. Sketch it in your booklet.
2. Carefully lower your egg into the cup.
3. Observe and sketch the level of water with the egg added.
4. Describe what happened.

What is the circumference of your egg?

1. Predict the circumference of your egg by cutting a piece of string you think will fit around your egg.
2. Check your prediction.
3. Was it too long? ...too short? ...just right?
4. Put your prediction on the graph near the station.
5. Use another piece of string and cut it the same size as the circumference of your egg.
6. Glue this string in your booklet.

Rolling Eggs

1. Predict where your egg will land when you roll it across the marked surface.
2. Place your egg marker on the space where you think it will land.
3. Roll your egg.
4. Place your egg marker on the graph.

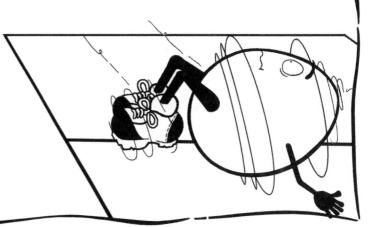

What do you think our eggs will look like after we dye them?

1. Place a spot of glue on your egg and press a small piece of fern into the glue.
2. Wrap a piece of nylon around your egg and close it with a twist tie.
3. Dip the egg into the dye.
4. Wait at least three minutes.
5. Remove the egg from the dye and let dry for one minute.
6. Unwrap the egg.
7. Compare your dyed egg with other eggs at the table.

What does the inside of an uncooked egg look like?

1. Crack a raw egg into the cup.
2. Draw what you observe in your booklet.
3. Do you see the following parts of the egg? Try to find them all.

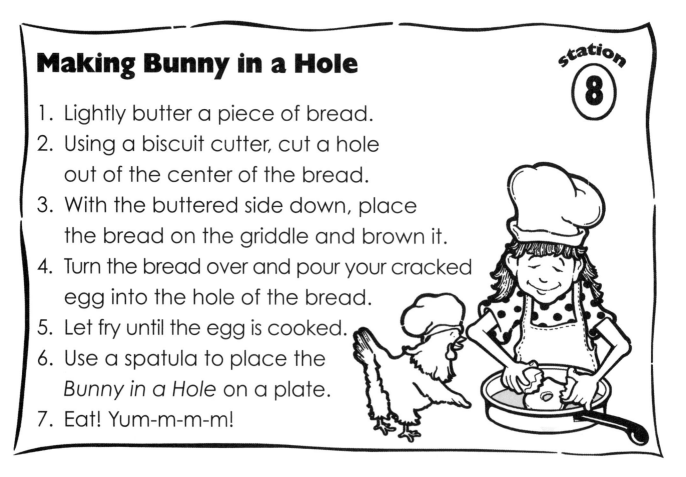

Membrane

Shell

Yolk

Albumen

Making Bunny in a Hole

1. Lightly butter a piece of bread.
2. Using a biscuit cutter, cut a hole out of the center of the bread.
3. With the buttered side down, place the bread on the griddle and brown it.
4. Turn the bread over and pour your cracked egg into the hole of the bread.
5. Let fry until the egg is cooked.
6. Use a spatula to place the *Bunny in a Hole* on a plate.
7. Eat! Yum-m-m-m!

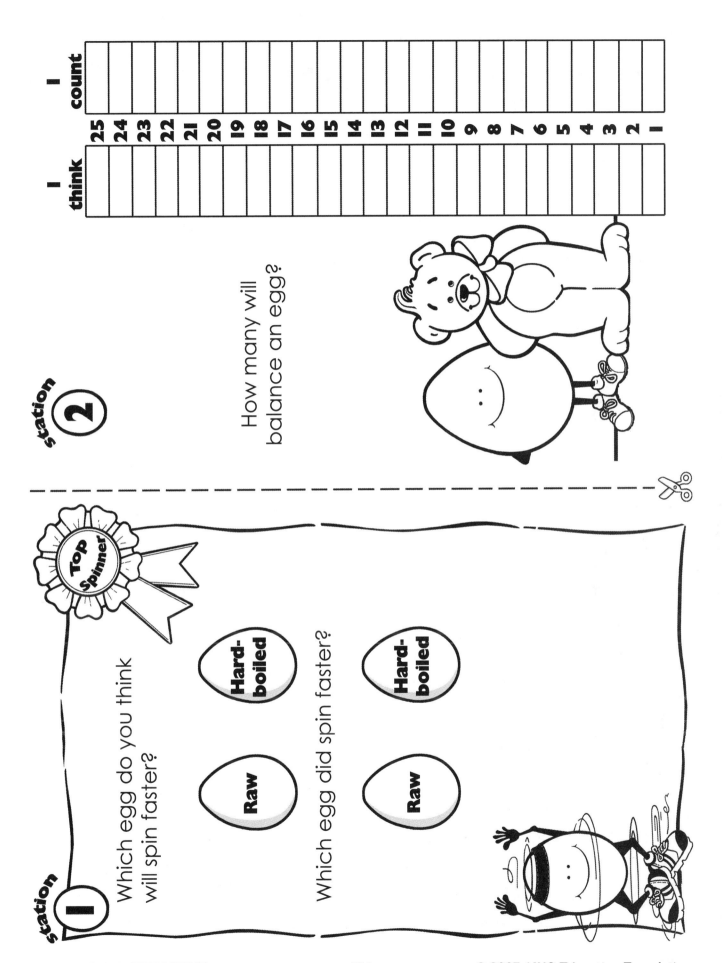

think | | | | | | | | | | | | | | | | | | | | | | | | |
count | | | | | | | | | | | | | | | | | | | | | | | | |

25 24 23 22 21 20 19 18 17 16 15 14 13 12 11 10 9 8 7 6 5 4 3 2 1

station 2

How many will balance an egg?

Top Spinner

station 1

Which egg do you think will spin faster?

Raw   Hard-boiled

Which egg did spin faster?

Raw   Hard-boiled

The circumference of my egg was

Draw the water level before adding your egg.

Draw the water level after adding your egg.

What happened?

132

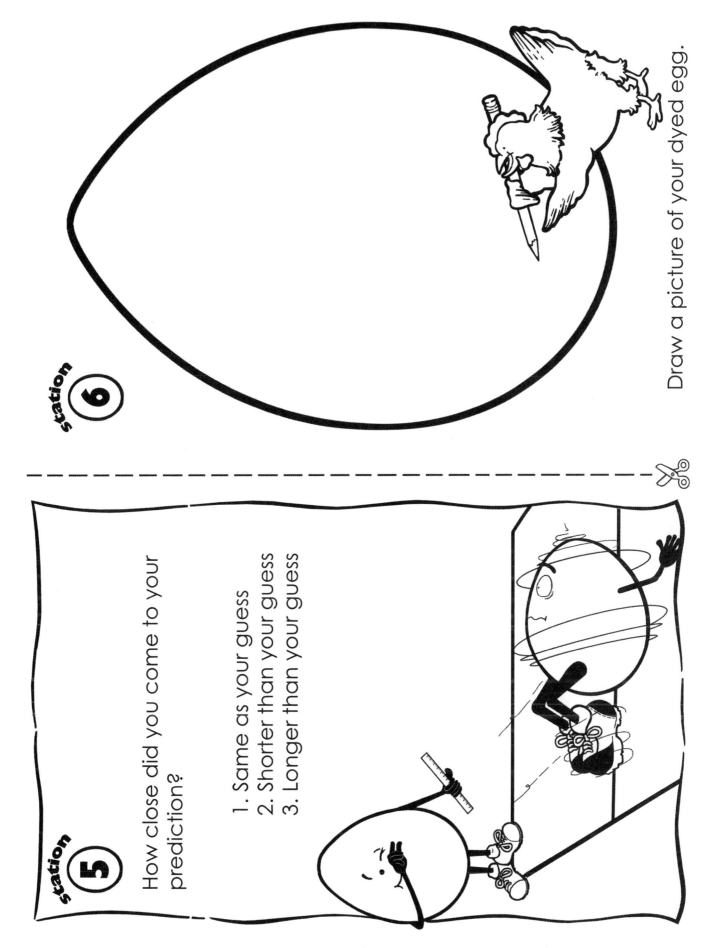

Draw a picture of your dyed egg.

How close did you come to your prediction?

1. Same as your guess
2. Shorter than your guess
3. Longer than your guess

station 8

Describe any changes you observe in the egg parts.

station 7

Draw the egg parts you see.

Cra-yolka crayons

Eggsploration Eggsploration Stations

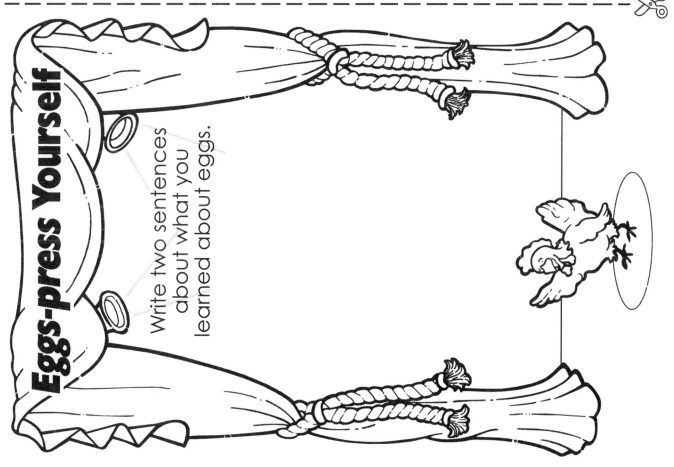

Eggs-press Yourself

Write two sentences about what you learned about eggs.

A Bear Eggs-pedition

Topic
Measurement

Key Question
How can you put plastic eggs in order from lightest to heaviest?

Learning Goals
Students will:
- order a set of eggs from lightest to heaviest using their hands,
- use a balance to find the actual order from lightest to heaviest, and
- find the mass of each egg in Teddy Bear Counters.

Guiding Documents
Project 2061 Benchmark
- *People use their senses to find out about their surroundings and themselves. Different senses give different information. Sometimes a person can get different information about the same thing by moving closer to it or further away from it.*

NRC Standard
- *Employ simple equipment and tools to gather data and extend the senses.*

NCTM Standards 2000*
- *Represent data using concrete objects, pictures, and graphs*
- *Apply and adapt a variety of appropriate strategies to solve problems*
- *Understand how to measure using nonstandard and standard units*
- *Use tools to measure*

Math
Measurement
   mass
Logical thinking
Graphing

Integrated Processes
Observing
Comparing and contrasting
Collecting and recording data
Predicting

Materials
For each group of students:
   four large plastic eggs (see *Management 2*)
   objects to fill the eggs
   balance
   Teddy Bear Counters or other counters

For each student:
   student pages

Background Information
Young students need multiple opportunities to investigate objects with different masses. Allowing time for making direct comparisons helps develop the ideas of heavier and lighter. To further develop these ideas, balances can be used for making the same type of comparisons. The teacher can then direct students to seeing the need to be more specific in describing the mass of objects by asking questions such as: How much heavier is this? How do you know? Non-customary units provide an interesting basis for this quantifying experience.

In this activity, students will be putting plastic eggs, each with a different mass, in order from lightest to heaviest using just the sense of touch. After making the initial prediction, they will use a balance to verify the prediction. After determining the order of the eggs, students will find the actual mass of each egg. The activity allows students to use a variety of problem-solving strategies and reinforces the idea of mass.

Management
1. This activity works best with small groups.
2. Prior to this activity, label the eggs in each set A, B, C, and D. Place different numbers of small objects such as pennies, Teddy Bear Counters, or BBs into each egg. Close and secure the eggs with transparent tape.
3. It is important that the students be able to feel some difference in mass between the eggs. Be sure that you select objects that allow you to have a range of masses.
4. If Teddy Bear Counters are not available to use as masses, centicubes, Friendly Bears, or other uniform objects can be substituted.
5. Before beginning the lesson, students should be given a time of free exploration to compare and find the mass of items using non-customary units.

Procedure

1. Ask the *Key Question*.
2. Tell the students that they will be looking at mystery eggs. The mystery is which egg has the most objects hiding inside, which has the fewest, and which are in between. The problem is that they cannot look inside the eggs to find out.
3. Have students get into groups and distribute a set of eggs to each group.
4. Ask the students to arrange all four eggs from lightest to heaviest just by lifting them. They will probably find it easiest to first identify the heaviest and lightest eggs, and then determine the placement of the remaining two eggs.
5. Distribute the first student page, and have students record the order they think the eggs go in from lightest to heaviest. Tell them that different people in the same group can have different answers.
6. Ask the students to predict what would happen if they put the heaviest egg in one pan of the balance and the lightest egg in the other pan. Distribute balances and have them test their predictions.
7. After the students have checked their predictions, have them record the correct order.
8. Ask the students if they can think of a way to determine the mass of the plastic eggs without looking inside.
9. If no student has suggested using a balance to find the mass of each egg, make the suggestion.
10. Have students use the balances to find the mass of each egg in Teddy Bear Counters.
11. Help them to record and graph their results.
12. After all groups have had a chance to complete the activity, gather the students together to discuss the results.

Connecting Learning

1. Were you able to put the eggs in the correct order from lightest to heaviest by lifting them? Why or why not?
2. Did everyone in your group have the same order with the eggs when they held them in their hands? Why or why not?
3. How did you use the balance to help you solve the problem?
4. Were the results different when you used the balance than when you held them in your hand? Why do you think there were differences?
5. What was the mass of the heaviest egg? What was the mass of the lightest egg?
6. Which two eggs had masses that were the closest together? Were these two eggs the hardest to tell apart? Why or why not?

*   Reprinted with permission from *Principles and Standards for School Mathematics*, 2000 by the National Council of Teachers of Mathematics. All rights reserved.

A Bear Eggs-pedition

When I hold the eggs in my hands, I think this is their order from lightest to heaviest.

Lightest

_____

_____

Heaviest

When I put the eggs in the balance, I know this is their order from lightest to heaviest.

Lightest

_____

_____

Heaviest

Honey

A Bear Eggs-pedition

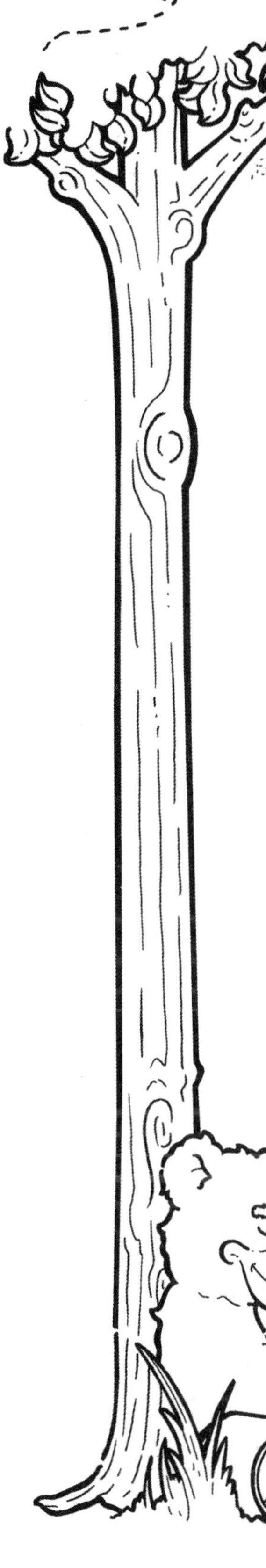

Find and record the mass of each egg.

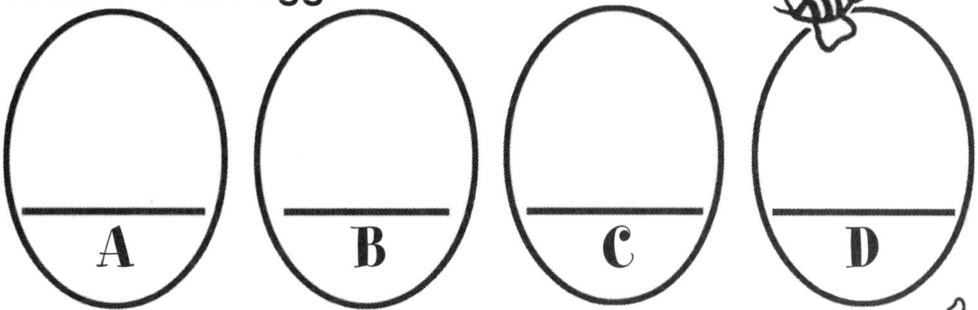

A   B   C   D

Graph your findings.

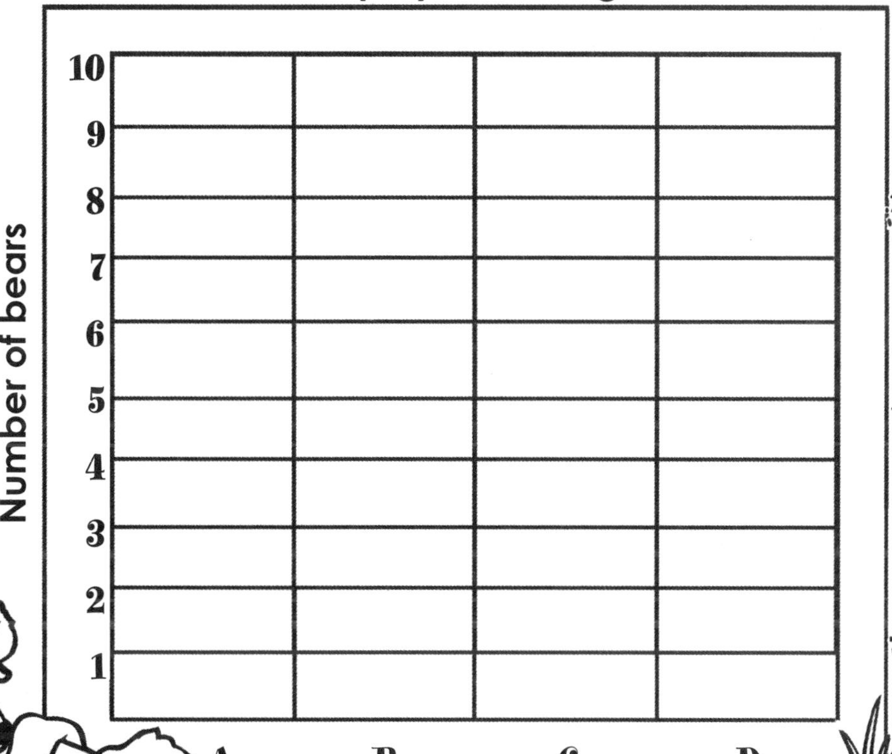

Number of bears

10
9
8
7
6
5
4
3
2
1

A   B   C   D

Egg

Which egg is lightest?

Which egg is heaviest?

Topic
Probability

Key Question
How many eggs of each color do you think are in the bag?

Learning Goals
Students will:
- record the results of repeated samples, and
- predict the number of each color of plastic egg in a bag based upon these samples.

Guiding Documents
Project 2061 Benchmark
- *Some things are more likely to happen than others. Some events can be predicted well and some cannot. Sometimes people aren't sure what will happen because they don't know everything that might be having an effect.*

NCTM Standards 2000*
- *Discuss events related to students' experiences as likely or unlikely*
- *Represent data using concrete objects, pictures, and graphs*
- *Create and use representations to organize, record, and communicate mathematical ideas*

Math
Probability
Counting
Logical thinking
Problem solving

Integrated Processes
Observing
Predicting
Collecting and recording data
Interpreting data
Generalizing

Materials
Paper grocery bag
10 colored plastic eggs (see *Management 1*)
Crayons
Student page

Background Information
Children's intuition can be developed through the study of probability at the primary level. By conducting experiments that draw upon children's experiences and interest, students make predictions and conduct experiments to test their predictions. As students conduct simple experiments and collect and analyze data, they look for patterns to help them predict probable outcomes. Children develop an understanding that some events are more likely to occur than others. They draw conclusions based upon their interpretations of data. They learn that it is possible to measure the likelihood of events. They see how larger samples of data give more reliable information than smaller sample sizes. Students find that an experiment may produce data that do not match theoretical probability.

In addition to developing an understanding of probability, this activity provides opportunities for students to collect, organize, and interpret data. Number sense, computational skills, thinking and reasoning skills are also developed.

Management
1. You will need at least 15 colored plastic eggs in two or three colors. To begin the activity, fill a paper grocery bag with 10 eggs—five each of two colors.
2. Each student will need crayons that match the colors of the eggs and multiple copies of the student page. (The number of copies will depend on the number of times you repeat the experience.)

Procedure
1. Discuss chance and probability with the class by discussing the terms *certain*, *likely*, *unlikely*, and *impossible*. Ask them to identify some things that fall into each of these categories. [sunrise, sun/rain tomorrow, snow in July, having superpowers, etc.]
2. Show the grocery bag to the class and explain that there are 10 colored eggs inside, but the number of colors and how many there are of each color is a mystery.
3. Distribute the student page and crayons and explain that everyone will be keeping track of what is pulled out of the bag.

4. Move around the classroom having different students draw one egg out of the bag. Have everyone record each egg's color by selecting the appropriate crayon and coloring in an egg on the recording page under *First Trial*. Direct the student who selected the egg to put it back into the bag.

5. Continue until the number of draws equals the total number of eggs—10.

6. Repeat steps four and five two more times.

7. Ask the students to look at their data and make a prediction about the number of each color of eggs in the bag. Have them color the appropriate eggs on the student page to reflect their predictions.

8. Use the first few *Connecting Learning* questions to help students think about the likely contents of the bag and what is possible, but unlikely.

9. Show them the eggs that are in the bag and have them color in the actual results and compare those to their predictions

10. Follow the same procedure, this time using a different number of each color. (Still use only two colors.) Discuss the results and compare them to the first experience.

11. Repeat the activity again as desired, using more than two colors of eggs. Discuss and compare the results to the previous experiences.

Connecting Learning

1. How many colors of eggs do you think there are in the bag? [two] Why?

2. Is it possible that there are more than two colors of eggs? [yes] Is it likely? [no] Why or why not? [We drew a total of 30 times and only saw two colors.]

3. How many of each color do you think are in the bag? Why do you think this?

4. Is it possible that there is only one ____ (color) egg in the bag? [yes] Is it likely? [no] Why or why not? [We saw the ____ (color) egg about as often as we saw the other color.]

5. How did what was actually in the bag compare to the samples we took? How did it compare to the prediction you made? Why?

6. What are some times that you need/want to know what is likely or unlikely to happen? [weather forecasts, etc.] How do you find the answers to your questions?

7. What are you wondering now?

Extensions

1. Use the concrete materials to develop or strengthen fractional concepts and ratios.

2. Use the activity to develop the concepts of greater than, less than, and equal to.

Home Link

Have students combine like objects of different colors to put in small bags to take home so they can do the activity with their families. Invite the students to share their families' results with the class.

* Reprinted with permission from *Principles and Standards for School Mathematics*, 2000 by the National Council of Teachers of Mathematics. All rights reserved.

First Trial

Second Trial

Third Trial

I guess the eggs are these colors.

The eggs are actually these colors.

The AIMS Program

AIMS is the acronym for "**A**ctivities **I**ntegrating **M**athematics and **S**cience." Such integration enriches learning and makes it meaningful and holistic. AIMS began as a project of Fresno Pacific University to integrate the study of mathematics and science in grades K-9, but has since expanded to include language arts, social studies, and other disciplines.

AIMS is a continuing program of the non-profit AIMS Education Foundation. It had its inception in a National Science Foundation funded program whose purpose was to explore the effectiveness of integrating mathematics and science. The project directors in cooperation with 80 elementary classroom teachers devoted two years to a thorough field-testing of the results and implications of integration.

The approach met with such positive results that the decision was made to launch a program to create instructional materials incorporating this concept. Despite the fact that thoughtful educators have long recommended an integrative approach, very little appropriate material was available in 1981 when the project began. A series of writing projects have ensued, and today the AIMS Education Foundation is committed to continue the creation of new integrated activities on a permanent basis.

The AIMS program is funded through the sale of books, products, and staff development workshops and through proceeds from the Foundation's endowment. All net income from program and products flows into a trust fund administered by the AIMS Education Foundation. Use of these funds is restricted to support of research, development, and publication of new materials. Writers donate all their rights to the Foundation to support its on-going program. No royalties are paid to the writers.

The rationale for integration lies in the fact that science, mathematics, language arts, social studies, etc., are integrally interwoven in the real world from which it follows that they should be similarly treated in the classroom where we are preparing students to live in that world. Teachers who use the AIMS program give enthusiastic endorsement to the effectiveness of this approach.

Science encompasses the art of questioning, investigating, hypothesizing, discovering, and communicating. Mathematics is the language that provides clarity, objectivity, and understanding. The language arts provide us powerful tools of communication. Many of the major contemporary societal issues stem from advancements in science and must be studied in the context of the social sciences. Therefore, it is timely that all of us take seriously a more holistic mode of educating our students. This goal motivates all who are associated with the AIMS Program. We invite you to join us in this effort.

Meaningful integration of knowledge is a major recommendation coming from the nation's professional science and mathematics associations. The American Association for the Advancement of Science in *Science for All Americans* strongly recommends the integration of mathematics, science, and technology. The National Council of Teachers of Mathematics places strong emphasis on applications of mathematics such as are found in science investigations. AIMS is fully aligned with these recommendations.

Extensive field testing of AIMS investigations confirms these beneficial results:

1. Mathematics becomes more meaningful, hence more useful, when it is applied to situations that interest students.
2. The extent to which science is studied and understood is increased, with a significant economy of time, when mathematics and science are integrated.
3. There is improved quality of learning and retention, supporting the thesis that learning which is meaningful and relevant is more effective.
4. Motivation and involvement are increased dramatically as students investigate real-world situations and participate actively in the process.

We invite you to become part of this classroom teacher movement by using an integrated approach to learning and sharing any suggestions you may have. The AIMS Program welcomes you!

Magazine

YOUR K-9 MATH AND SCIENCE
CLASSROOM ACTIVITIES RESOURCE

The AIMS Magazine is your source for standards-based, hands-on math and science investigations. Each issue is filled with teacher-friendly, ready-to-use activities that engage students in meaningful learning.

• *Four issues each year (fall, winter, spring, and summer).*

Current issue is shipped with all past issues within that volume.

| 1821 | Volume XXI | 2006-2007 | $19.95 |
| 1822 | Volume XXII | 2007-2008 | $19.95 |

Two-Volume Combination
| M20507 | Volumes XX & XXI | 2005-2007 | $34.95 |
| M20608 | Volumes XXI & XXII | 2006-2008 | $34.95 |

Back Volumes Available
Complete volumes available for purchase:

1802	Volume II	1987-1988	$19.95
1804	Volume IV	1989-1990	$19.95
1805	Volume V	1990-1991	$19.95
1807	Volume VII	1992-1993	$19.95
1808	Volume VIII	1993-1994	$19.95
1809	Volume IX	1994-1995	$19.95
1810	Volume X	1995-1996	$19.95
1811	Volume XI	1996-1997	$19.95
1812	Volume XII	1997-1998	$19.95
1813	Volume XIII	1998-1999	$19.95
1814	Volume XIV	1999-2000	$19.95
1815	Volume XV	2000-2001	$19.95
1816	Volume XVI	2001-2002	$19.95
1817	Volume XVII	2002-2003	$19.95
1818	Volume XVIII	2003-2004	$19.95
1819	Volume XIX	2004-2005	$19.95
1820	Volume XX	2005-2006	$19.95

Volumes II to XIX include 10 issues.

Call 1.888.733.2467 or
go to www.aimsedu.org

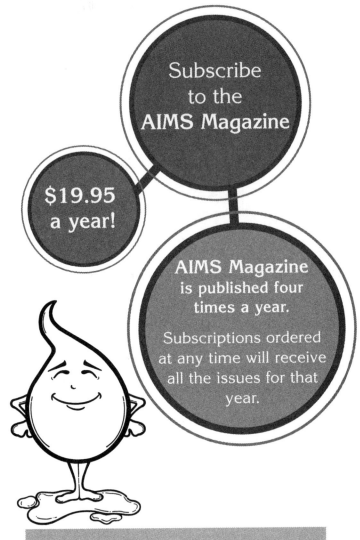

Subscribe to the AIMS Magazine

$19.95 a year!

AIMS Magazine is published four times a year.

Subscriptions ordered at any time will receive all the issues for that year.

AIMS Online—www.aimsedu.org

To see all that AIMS has to offer, check us out on the Internet at www.aimsedu.org. At our website you can search our activities database; preview and purchase individual AIMS activities; learn about core curriculum, college courses, and workshops; buy manipulatives and other classroom resources; and download free resources including articles, puzzles, and sample AIMS activities.

AIMS News
While visiting the AIMS website, sign up for AIMS News, our FREE e-mail newsletter. You'll get the latest information on what's new at AIMS including:

• New publications;
• New core curriculum modules; and
• New materials.

Sign up today!

Duplication Rights

Standard
Duplication
Rights

Purchasers of AIMS activities (individually or in books and magazines) may make up to 200 copies of any portion of the purchased activities, provided these copies will be used for educational purposes and only at one school site.

Workshop or conference presenters may make one copy of a purchased activity for each participant, with a limit of five activities per workshop or conference session.

Standard duplication rights apply to activities received at workshops, free sample activities provided by AIMS, and activities received by conference participants.

All copies must bear the AIMS Education Foundation copyright information.

Unlimited
Duplication
Rights

To ensure compliance with copyright regulations, AIMS users may upgrade from standard to unlimited duplication rights. Such rights permit unlimited duplication of purchased activities (including revisions) for use at a given school site.

Activities received at workshops are eligible for upgrade from standard to unlimited duplication rights.

Free sample activities and activities received as a conference participant are not eligible for upgrade from standard to unlimited duplication rights.

Upgrade
Fees

The fees for upgrading from standard to unlimited duplication rights are:
- $5 per activity per site,
- $25 per book per site, and
- $10 per magazine issue per site.

The cost of upgrading is shown in the following examples:
- activity: 5 activities x 5 sites x $5 = $125
- book: 10 books x 5 sites x $25 = $1250
- magazine issue: 1 issue x 5 sites x $10 = $50

Purchasing
Unlimited
Duplication
Rights

To purchase unlimited duplication rights, please provide us the following:
1. The name of the individual responsible for coordinating the purchase of duplication rights.
2. The title of each book, activity, and magazine issue to be covered.
3. The number of school sites and name of each site for which rights are being purchased.
4. Payment (check, purchase order, credit card)

Requested duplication rights are automatically authorized with payment. The individual responsible for coordinating the purchase of duplication rights will be sent a certificate verifying the purchase.

Internet
Use

Permission to make AIMS activities available on the Internet is determined on a case-by-case basis.

• P. O. Box 8120, Fresno, CA 93747-8120 •
• aimsed@aimsedu.org • www.aimsedu.org •
• 559.255.6396 (fax) • 888.733.2467 (toll free) •

AIMS Program Publications

Actions with Fractions, 4-9
Awesome Addition and Super Subtraction, 2-3
Bats Incredible! 2-4
Brick Layers II, 4-9
Chemistry Matters, 4-7
Counting on Coins, K-2
Cycles of Knowing and Growing, 1-3
Crazy about Cotton, 3-7
Critters, 2-5
Electrical Connections, 4-9
Exploring Environments, K-6
Fabulous Fractions, 3-6
Fall into Math and Science, K-1
Field Detectives, 3-6
Finding Your Bearings, 4-9
Floaters and Sinkers, 5-9
From Head to Toe, 5-9
Fun with Foods, 5-9
Glide into Winter with Math and Science, K-1
Gravity Rules! 5-12
Hardhatting in a Geo-World, 3-5
It's About Time, K-2
It Must Be A Bird, Pre-K-2
Jaw Breakers and Heart Thumpers, 3-5
Looking at Geometry, 6-9
Looking at Lines, 6-9
Machine Shop, 5-9
Magnificent Microworld Adventures, 5-9
Marvelous Multiplication and Dazzling Division, 4-5
Math + Science, A Solution, 5-9
Mostly Magnets, 2-8
Movie Math Mania, 6-9
Multiplication the Algebra Way, 6-8
Off the Wall Science, 3-9
Out of This World, 4-8
Paper Square Geometry:
    The Mathematics of Origami, 5-12
Puzzle Play, 4-8
Pieces and Patterns, 5-9
Popping With Power, 3-5
Positive vs. Negative, 6-9
Primarily Bears, K-6
Primarily Earth, K-3
Primarily Physics, K-3
Primarily Plants, K-3

Problem Solving: Just for the Fun of It! 4-9
Problem Solving: Just for the Fun of It! Book Two, 4-9
Proportional Reasoning, 6-9
Ray's Reflections, 4-8
Sensational Springtime, K-2
Sense-Able Science, K-1
Soap Films and Bubbles, 4-9
Solve It! K-1: Problem-Solving Strategies, K-1
Solve It! 2nd: Problem-Solving Strategies, 2
Solve It! 3rd: Problem-Solving Strategies, 3
Solve It! 4th: Problem-Solving Strategies, 4
Solve It! 5th: Problem-Solving Strategies, 5
Spatial Visualization, 4-9
Spills and Ripples, 5-12
Spring into Math and Science, K-1
The Amazing Circle, 4-9
The Budding Botanist, 3-6
The Sky's the Limit, 5-9
Through the Eyes of the Explorers, 5-9
Under Construction, K-2
Water Precious Water, 2-6
Weather Sense: Temperature, Air Pressure, and Wind, 4-5
Weather Sense: Moisture, 4-5
Winter Wonders, K-2

Spanish Supplements*
Fall Into Math and Science, K-1
Glide Into Winter with Math and Science, K-1
Mostly Magnets, 2-8
Pieces and Patterns, 5-9
Primarily Bears, K-6
Primarily Physics, K-3
Sense-Able Science, K-1
Spring Into Math and Science, K-1

* Spanish supplements are only available as downloads from the
  AIMS website. The supplements contain only the student pages
  in Spanish; you will need the English version of the book for the
  teacher's text.

Spanish Edition
Constructores II: Ingeniería Creativa Con Construcciones
    LEGO® 4-9
    The entire book is written in Spanish. English pages not included.

Other Publications
Historical Connections in Mathematics, Vol. I, 5-9
Historical Connections in Mathematics, Vol. II, 5-9
Historical Connections in Mathematics, Vol. III, 5-9
Mathematicians are People, Too
Mathematicians are People, Too, Vol. II
What's Next, Volume 1, 4-12
What's Next, Volume 2, 4-12
What's Next, Volume 3, 4-12

For further information write to:
AIMS Education Foundation • P.O. Box 8120 • Fresno, California 93747-8120
www.aimsedu.org • 559.255.6396 (fax) • 888.733.2467 (toll free)